Springer Tracts in Advanced Robotics 125

Series editors

More information about this series at http://www.springer.com/series/5208

Angel Santamaria-Navarro
Joan Solà · Juan Andrade-Cetto

Visual Guidance of Unmanned Aerial Manipulators

Springer

Angel Santamaria-Navarro
Institut de Robòtica i Informàtica Industrial, CSIC-UPC
Barcelona, Spain

Juan Andrade-Cetto
Institut de Robòtica i Informàtica Industrial, CSIC-UPC
Barcelona, Spain

Joan Solà
Institut de Robòtica i Informàtica Industrial, CSIC-UPC
Barcelona, Spain

ISSN 1610-7438 ISSN 1610-742X (electronic)
Springer Tracts in Advanced Robotics
ISBN 978-3-030-07217-9 ISBN 978-3-319-96580-2 (eBook)
https://doi.org/10.1007/978-3-319-96580-2

This Springer imprint is published by the registered company Springer Nature Switzerland AG
The registered company address is: Gewerbestrasse 11, 6330 Cham, Switzerland

To Núria and my family

Foreword

Robotics is undergoing a major transformation in scope and dimension. From a largely dominant industrial focus, robotics is rapidly expanding into human environments and vigorously engages in its new challenges. Interacting with, assisting, serving, and exploring with humans, the emerging robots will increasingly touch people and their lives.

Beyond its impact on physical robots, the body of knowledge robotics has produced is revealing a much wider range of applications reaching across diverse research areas and scientific disciplines, such as biomechanics, haptics, neurosciences, virtual simulation, animation, surgery, and sensor networks, among others. In return, the challenges of the new emerging areas are proving an abundant source of stimulation and insights for the field of robotics. It is indeed at the intersection of disciplines that the most striking advances happen.

The *Springer Tracts in Advanced Robotics (STAR)* is devoted to bringing to the research community the latest advances in the robotics field on the basis of their significance and quality. Through a wide and timely dissemination of critical research developments in robotics, our objective with this series is to promote more exchanges and collaborations among the researchers in the community and contribute to further advancements in this rapidly growing field.

The monograph by Angel Santamaria-Navarro, Joan Solà, and Juan Andrade-Cetto is based on the first author's doctoral thesis. It deals with the hot research topic of visual guidance of unmanned aerial manipulators, which can be found in several industrial inspection and maintenance scenarios. A full coverage of an autonomous navigation technique, a guidance method using visual information, and a hierarchical control algorithm to handle multiple-priority tasks with redundancy resolution are provided, and how these are effective for aerial manipulators interacting with the environment is keenly discussed with supporting practical issues.

Rich of examples developed by means of extensive experimentation on real robotic platforms, this volume was a finalist for the 2018 Georges Giralt Ph.D. Award. A very fine addition to the STAR series!

Naples, Italy
June 2018

Bruno Siciliano
STAR Editor

Preface

The ability to fly has greatly expanded the possibilities for robots to perform surveillance, inspection, or map generation tasks. Yet, it was only in recent years that research in aerial robotics was mature enough to allow active interactions with the environment. The robots responsible for these interactions are called aerial manipulators and usually combine a multirotor platform and one or more robotic arms.

The main objective of this book is to formalize the concept of aerial manipulator and present guidance methods, using visual information, to provide them with autonomous functionalities.

A key competence to control an aerial manipulator is the ability to localize it in the environment. Traditionally, this localization has required an external infrastructure of sensors (GPS or IR cameras), restricting the real applications. Furthermore, localization methods with onboard sensors, exported from other robotics fields such as simultaneous localization and mapping (SLAM), require large computational units becoming a handicap in vehicles where size, load, and power consumption are important restrictions. In this regard, this book proposes a method to estimate the state of the vehicle (position, orientation, velocity, and acceleration) by means of onboard, low-cost, lightweight, and high-rate sensors.

With the physical complexity of these robots, it is required to use advanced control techniques during navigation. Thanks to their redundancy on degrees-of-freedom, they offer the possibility to accomplish not only with mobility requirements but also with other tasks simultaneously and hierarchically, prioritizing them depending on their impact on the overall mission success. In this book, we present such control laws and define a number of these tasks to drive the vehicle using visual information, guarantee the robot integrity during flight, and improve the platform stability or increase arm operability.

The main contributions of this research work are threefold: (1) Present a localization technique to allow autonomous navigation; this method is specifically designed for aerial platforms with size, load, and computational burden restrictions. (2) Obtain control commands to drive the vehicle using visual information (visual servo). (3) Integrate the visual servo commands into a hierarchical control law by exploiting the redundancy of the robot to accomplish secondary tasks during flight. These tasks are specific for aerial manipulators, and they are also provided.

All the techniques presented in this book have been validated throughout extensive experimentation with real robotic platforms.

Barcelona, Spain
May 2018

Angel Santamaria-Navarro
Joan Solà
Juan Andrade-Cetto

Contents

Acronyms

$\mathcal{SE}(3)$	Special Euclidean group
$\mathcal{SO}(3)$	Special orthogonal group
$\mathfrak{so}(3)$	Tangent space to $\mathcal{SO}(3)$
AEROARMS	EU project: Aerial robotics system integrating multiple arms and advanced manipulation capabilities for inspection and maintenance (H2020-ICT-2014-1-644271)
AEROWORKS	EU project: Collaborative aerial robotic workers (H2020-ICT-2014-1-644128)
ANEES	Average normalized estimation error
ARCAS	EU project: Aerial robotics cooperative assembly system (FP7-ICT-287617)
CoG	Center of gravity
CPU	Central processing unit
DoF	Degrees-of-freedom
EKF	Extended Kalman filter
EPnP	Efficient perspective-n-point
ESKF	Error-state Kalman filter
FoV	Field of view
GE	Globally defined error
GPS	Global positioning system
HTPC	Hierarchical task priority control
HVS	Hybrid visual servo
IBVS	Image-based visual servo
IMU	Inertial measurement unit
IR	Infrared
LE	Locally defined error
LIDAR	Light detection and ranging
LLS	Local least squares
MAV	Micro aerial vehicle
MSKF	Multi-state Kalman filter

NEES	Normalized estimation error
PBVS	Position-based visual servo
PERCH	Penn Engineering Research Collaborative Hub
Q0B	Quaternion backward zeroth-order integration
Q0F	Quaternion forward zeroth-order integration
Q1	Quaternion first-order integration
QP	Quadratic programming
RADAR	Radio detection and ranging
RF	Radio frequency
RLS	Recursive least squares
ROS	Robot operating system
SLAM	Simultaneous localization and mapping
STD	Standard deviation
UAM	Unmanned aerial manipulator
UAV	Unmanned aerial vehicle
UIBVS	Uncalibrated image-based visual servo
UPnP	Uncalibrated perspective-n-point
VIO	Visual-inertial odometry
VI-SLAM	Visual-inertial localization and mapping
VS	Visual servo
VToL	Vertical take-off and landing

List of Figures

List of Tables

Chapter 1
Introduction

Human beings have always tried to surpass their own capabilities. This overcoming instinct and the continuous evolution of technology have led, at the beginning of the last century, to important advances in the fields of robotics and aeronautics, to such a degree that the ability to fly became not only a skill reserved to the animal kingdom, but a feasible, reliable and even needed type of transport in modern societies. While commercial planes became part of daily life, the development of other types of aerial vehicles also experienced important boosts, e.g. multirotors. However, all these apparatus had an important restriction: their control required complicated abilities, thus in all designs human pilots were required to operate from aboard. It was not until the end of the twentieth century that technology offered the possibility to substitute the human pilot by automatic control systems, leading to the first Unmanned Aerial Vehicles (UAV) without the need to carry the load of human pilots. These UAVs were more economical, small and light-weight than manned aerial vehicles, and provided the opportunity to perform a new bunch of tasks without compromising pilot lives.

UAVs, and in particular multirotor systems, have substantially gained popularity in recent years, motivated by their significant increase in maneuverability, together with a decrease in weight and cost [1, 8]. UAVs are not usually required to interact physically with the environment, but only to perform tracking, surveillance or inspection tasks. Applications however are now appearing for cases in which physical interaction is needed. Some examples are ARCAS, AEROARMS and AEROWORKS[1] EU funded projects with the aim to develop UAV systems with advanced manipulation capabilities for autonomous industrial inspection and repair tasks. This new type of aerial vehicles are named Unmanned Aerial Manipulators (UAM) and consist of a multirotor platform (able to navigate, hover in position or take-off and land vertically) with one or more robotic manipulators, usually endowed below, to interact with the environment.

This book is focused on the navigation of UAMs by means of visual information, entailing methods for robot state estimation (i.e., to obtain its position, orientation,

[1] www.aeroarms-project.eu, www.aeroworks2020.eu.

A. Santamaria-Navarro et al., *Visual Guidance of Unmanned Aerial Manipulators*, Springer Tracts in Advanced Robotics 125,
https://doi.org/10.1007/978-3-319-96580-2_1

velocity and acceleration), and task oriented control. The motivation leading to this type of research is presented next, followed by the specific objectives and scope of the book, and an outline of the approach taken to meet such objectives, concluding with the scientific publications product of the work.

There exist a number of situations where humans must but cannot operate. Dangerous or hazardous environments (e.g., zones contaminated with radiation) are clear examples because human lives are at risk. In some other tasks, like inspection and maintenance in the industry, human operations are with acceptable risks but carry large economic costs (e.g., repair of high and inaccessible machinery parts). In most of those cases, deploying an autonomous UAM with the ability to fly and manipulate is a reasonable solution.

UAMs are expected to navigate and move efficiently on different 3D scenarios depending on their assigned tasks. With this purpose, several perception techniques have been developed in recent years, such as those relying on external infrastructure, e.g. motion capture systems [3], GPS [5] or RF beacons [4], which may not always be practical for UAM operations, specially outdoors. Alternatively, there exist a number of methods embarking all hardware and software for self-localization, e.g. [1], however they generally require powerful computers. Regarding UAMs, where restrictions like size, payload or computational burden are challenging, it is interesting the use of onboard, small, and light-weight sensors such as cameras or inertial measurement units (IMU) to drive autonomously the robot. The first part of this book is focused in solving such perception challenges for accurate localization and guidance of UAMs in GPS-denied environments, distinguishing between navigation operations where a coarse localization (i.e., state estimation) suffices to drive the vehicle, and manipulation tasks where visual servoing techniques must be applied to reach the required localization precision.

The use of multirotors, specifically small-sized models called Micro Aerial Vehicles (MAVs), have substantially gained popularity in the research community in recent years as UAM flying platforms, motivated by the significant increase in maneuverability, together with a decrease in weight and cost. The ability for MAVs to manipulate or carry objects greatly expanded the types of missions achievable by such unmanned systems. High performance arms typically weigh more than 10 kg and cannot be supported by most commercially available small-sized multirotors. However, recent developments suggest a trend change with platform payload capabilities increasing and arm weights getting smaller [2, 6].

Flying with a suspended load is a challenging task since the vehicle is characterized by unstable dynamics in which the presence of the object causes nontrivial coupling effects and the load significantly changes the flight characteristics. Given that the stability of the vehicle-load system must be preserved, it is essential for the flying robot to have the ability to minimize the effects of the arm in the flying system during the assigned maneuvers [7]. Moreover, multirotors are usually equipped with four, six or eight aligned coplanar propellers and, due to their symmetric design, motion control is achieved by altering the rotation rate of one or more of these propellers, thereby changing its torque load and thrust lift characteristics. With these actuation techniques, multirotors become under-actuated vehicles with only 4 degrees-of-freedom (DoF) at a high-level of control (e.g., one for the thrust and three torques), a restriction that affects the manipulator base movements and thus must be considered during flight. In the second part of this book we explore and present techniques to reduce such platform control restrictions while reducing the coupling effects of endowing a light-weight serial arm.

During UAM missions we can distinguish two main navigation phases. First, the approaching and exiting maneuvers were the vehicle is expected to navigate to, or from, a point where the manipulation target is close enough to start interaction, or at least at sight. In a first instance, this navigation requires the knowledge of our own position and orientation in the environment. Hence, vehicle's state estimation is one of the keys in this work. State estimation for small-sized and light-weight platforms is a challenging task, due to the limited computing capacity that these vehicles can carry, greatly reducing the admissible complexity of the onboard algorithms, and the limited payload, greatly restricting both the weight and the physical dimensions of onboard sensors. Our objective is to develop a state estimation method to compute the odometry of the flying vehicle by means of light-weight sensors which can work at a high rate requiring low computational burden.

A second navigation phase is determined for those situations where the robot should be driven precisely to perform manipulation tasks. Physical interaction with the environment calls for positioning accuracy at the centimeter level, which in GPS denied environments is often difficult to achieve. In this regard, the use of vision (i.e., visual servo) is a widely adopted solution to cope with unknown environments. The development of such visual servo techniques is another purpose of this book.

In both navigation phases, flying with a suspended load is a challenging operation due to undesired dynamic effects of the coupled bodies. But, as shown in this work, the attachment of a manipulator arm to the base of the flying platform can be seen as a strategy to alleviate these effects. The arm lets us exploit the redundancy on DoFs of the overall system not only to achieve the desired guidance tasks, but to do so whilst attaining also other tasks during the mission, compensating for changes in weight distribution during arm operation and driving the arm to a desired configuration with high manipulability, thus improving overall flight behavior.

In summary, the main goal of this book is to provide UAMs with autonomous navigation capabilities. The necessary objectives are listed as follows:

- Define UAM systems and characteristics regarding navigation and control tasks.
- Study odometry estimation methods, in terms of simple and light algorithms, using light-weight, high-rate and low power-consuming sensors.
- Define visual servo techniques to be used within UAM settings.
- Integrate visual guidance operations in a control law considering a hierarchical task composition.
- Define specific tasks to alleviate undesired dynamic effects of the coupled bodies.
- Explore how the non-controllable DoFs of the platform must be considered in the task designs and the hierarchical control law.
- Validate the presented techniques with simulation case studies and exhaustive experimentation.

References

1. Blösch, Michael, Stephan Weiss, Davide Scaramuzza, and Roland Siegwart. 2010. Vision based MAV navigation in unknown and unstructured environments. In *IEEE international conference on robotics and automation*, pp. 21–28, Anchorage, USA.
2. Korpela, C.M., T.W. Danko, and P.Y. Oh. 2011. Designing a system for mobile manipulation from an unmanned aerial vehicle. In *IEEE conference on technologies for practical robot applications*, pp. 109–114, Woburn, USA.
3. Liu, Hui, Houshang Darabi, Pat Banerjee, and Jing Liu. 2007. Survey of wireless indoor positioning techniques and systems. *IEEE Transactions on Systems, Man, and Cybernetics* 37 (6): 1067–1080.
4. Mueller, Mark W., Michael Hamer, and Raffaello D'Andrea. 2015. Fusing ultra-wideband range measurements with accelerometers and rate gyroscopes for quadrocopter state estimation. In *IEEE international conference on robotics and automation*, pp. 1730–1736, Seattle, USA.
5. Nemra, Abdelkrim, and Nabil Aouf. 2010. Robust INS/GPS sensor fusion for UAV localization using SDRE nonlinear filtering. *IEEE Sensors Journal* 10 (4): 789–798.
6. Orsag, Matko, Christopher Korpela, and Oh Paul. 2013. Modeling and control of MM-UAV: Mobile manipulating unmanned aerial vehicle. *Journal of Intelligent & Robotic Systems* 69 (1–4): 227–240.
7. Palunko, I., P. Cruz, and R. Fierro. 2012. Agile load transportation: Safe and efficient load manipulation with aerial robots. *IEEE Robotics Automation Magazine* 19 (3): 69–79.
8. Tomiá, Teodor, Korbinian Schmid, Philipp Lutz, Andreas Dömel, Michael Kassecker, Elmar Mair, Iris Lynne Grixa, Felix Ruess, Michael Suppa, and Darius Burschka. 2012. Toward a fully autonomous UAV: Research platform for indoor and outdoor urban search and rescue. *IEEE Robotics & Automation Magazine* 19 (3): 46–56.

Chapter 2
Robot State Estimation

2.1 Introduction

Micro aerial vehicles have gained significant attention in the last decade both in academia and industry, mostly due to their potential use in a wide range of applications such as exploration [46], inspection [52], mapping, interaction with the environment [7, 44], search and rescue [27], and to their significant mechanical simplicity and ease of control. Moreover, their ability to operate in confined spaces, hover in space and even perch, together with a decrease in cost make them very attractive with tremendous potential as UAM flying platforms.

MAVs are in general nonlinear and highly unstable systems and a clever combination of sensors and controllers must be devised to fly them autonomously in a robust and efficient way. Most of the approaches model the aerial platform as two connected ideal subsystems and use a nested control structure:

- A low-level attitude controller, usually running at 1 kHz and using the IMU readings as an inclinometer to estimate the platform tilt.
- A high-level controller for global 3D positioning of the vehicle, usually running below 50 Hz and requiring more complex sensor suites.

With this nested scheme, the attitude controller has to be faster than the position controller because of the vehicle's high dynamics. To close these control loops and enable autonomous missions with MAVs, a robust, accurate and high update rate state estimation pipeline is crucial.

In this chapter we present a fast and low-cost state estimation method for small-sized multirotors. We use exclusively low-cost and high-rate sensors together with low-complexity algorithms to allow its use on board in computational units with low processing capacity.

As software, we have developed two Kalman filters, in the extended (EKF) and error-state (ESKF) forms [32], together with a wide range of variations in the inner details, for the sake of performance evaluation and comparison. To show the feasi-

A. Santamaria-Navarro et al., *Visual Guidance of Unmanned Aerial Manipulators*, Springer Tracts in Advanced Robotics 125,
https://doi.org/10.1007/978-3-319-96580-2_2

bility of the proposed estimation pipeline, a non-linear control law is also presented to drive autonomously aerial platforms using the estimated state. This controller is fed with trajectories generated with a planning approach also defined here.

To the knowledge of the authors, the papers related to this chapter, [37, 38], were the first examples of usage of such a low-cost flow-range-inertial sensor setup for 6 DoF motion estimation, using also low complexity algorithms without sophisticated features. This sensor setup has the advantage of being simple, light-weight, low-cost and requiring low power-consumption, and it is already included, with minor variations, in several commercial multirotors as their basic instrumentation, being typically used as a means for automatic hovering.

This chapter is organized as follows. The next Section gives an overview of the state-of-art on MAV localization systems. Basic concepts and notations are provided in Sect. 2.3. Section 2.4 presents the sensor suite with the corresponding observation models. The system kinematics are defined in Sect. 2.5. Sections 2.6 and 2.7 reveal the main filtering details. A non-linear control law and a dynamically feasible trajectory planning approaches are presented in Sect. 2.9. Section 2.10 contains simulations and experimental results that validate the proposed methods. Finally, Sect. 2.11 concludes the work and provides a summary of main contributions.

2.2 Related Work

Low-cost, low complexity solutions for MAV state estimation are not very common. A first family of localization systems relies on external infrastructure using GPS, RF beacons, visual tags or infrared cameras, e.g. [19], which may not always be practical. When using low-cost setups, these systems usually lack precision, dynamics, or both. A second family of methods embark all hardware and software for self-localization, thus not relying on any external setup. Assuming that a map is not available in advance, e.g.[51] or [18], the self-localization approaches include any kind of simultaneous localization and mapping (SLAM) or odometry systems drawn from other robotics fields [5]. Good results have been obtained using stereo camera configurations [12, 41, 46], and RGB-D sensor systems [21, 28, 39, 48]. However, these algorithms generally require powerful computers to produce and deal with fairly dense point clouds. Moreover, RGB-D sensors have low quality cameras and suffer during exposure to direct sunlight.

State estimation for small-sized and light-weight platforms is a challenging task, due to the limited computing capacity that these vehicles can carry, greatly reducing the admissible complexity of the on-board algorithms, and the limited payload, restricting both the weight and physical dimensions of the on-board sensors. Thus, solutions requiring bulky sensors like RADARS or LIDARS are not yet an option.

Not surprisingly, combinations of IMU and monocular visual sensors are becoming very popular, thanks to their low weight, power consumption and cost, and their ease of installation. This constitutes a minimalist yet powerful sensor suite for autonomous localization, as it allows recovering both the high motion dynamics and the localization with respect to the environment, including scale and, most important for MAV navigation, the direction of gravity [11, 25].

The processes of estimating the vehicle state using such sensors are known as visual-inertial odometry (VIO, with no absolute localization) [9, 17, 33, 40], and visual-inertial SLAM (VI-SLAM, enabling absolute localization by revisiting previously mapped areas) [3, 8, 13, 34, 50]. The focus of our work is not at building maps, and we concentrate on VIO.

There exist a multitude of VIO approaches. Roumeliotis et al. [33] obtain relative pose measurements of a vehicle using two-frame feature based motion estimation and then augment the information with IMU measurements. More recent approaches, [17, 29], implement the Multi-State Kalman Filter (MSKF), which estimates a sliding window of camera poses in the filter state and uses feature measurements to impose probabilistic constraints on the camera poses.

In [14] a system is developed using a bundle adjustment technique where each sensory system produces a pose estimate which gets fused with other pose estimates in an EKF formulation. The pose estimate is sub-optimal since the existing cross-correlations between internal states of the different devices are unused. Iterative minimization on a window of poses within a filter using graphical models is presented in [22]. As new poses are added and past observations are removed, the current window of states being estimated is independent from the previous estimates with the advantage that errors are isolated to the region where they occur.

The high precision outcome in most of the above-mentioned VIO methods is attained by jointly estimating a subset of past camera poses and a number of landmarks in the environment, which are tracked in the image over relatively long periods and require the estimation of a very large state vector. These tracks are exploited either through the explicit representation of 3D points in the form of landmarks in the map, or through the proper exploitation of the induced epipolar constraints, creating a graph of constraints between vehicle states and landmarks, which is incrementally solved by nonlinear optimization. These VIO solutions are computationally intensive, and require carefully optimized code to compensate for the limited computing resources typical of MAVs.

To reduce the computational burden and to increase the update rate, several authors opt to exploit the image information only locally and in 2D. Weiss et al. [49] proposes a speed-estimation module, which converts the camera-IMU pair into a metric body-speed sensor at a 40 Hz update rate, using optical flow information of at least 2 image features, within an EKF framework. A parallel tracking and mapping (PTAM) pipeline is tailored to the system, which therefore requires on-board image processing. In [31], flow information is fused with inertial measurements. However, only simulation results are provided. A similar approach [2] presents a novel visual error term and uses a visual-inertial sensor consisting on a synchronized global-shutter camera and IMU [30] to obtain flow information, though running at 20 Hz. In this Chapter we present a similar method to fuse IMU and flow readings but using a smart camera. This smart camera computes internally all image processing required to obtain the optical flow, thus allowing a fast and simple algorithmic solution for vehicle state estimation running on the main on-board computational unit.

We consider this combination of IMU and monocular visual sensors a minimal sensor suite for autonomous localization because not all dimensions in the state are observable, e.g. [9, 17, 24, 25]. The observability analysis proves the theoretical observability of *roll* and *pitch* of the vehicle, and intrinsic sensor calibration states (IMU biases). The extrinsic calibration (camera-to-IMU transformation) and the map scale are observable in the system if tracking of image features are done over time. As we exploit the image information only locally and in 2D we take advantage of a range sensor to obtain the scale, and therefore, the speed of the vehicle. We aim at tracking position, velocity and orientation using an IMU, a smart camera and a range sensor (either a sonar or an infrared time-of-flight sensors). As stated later in Sect. 2.8 x, y, and *yaw* will not be observable and thus subject to drift over time.

2.3 Basic Concepts and Notation

We consider a quadrotor, equipped with an IMU, a flow smart camera and a range sensor, moving with respect to a global coordinate frame assumed inertial. All sensors are rigidly attached together below the aerial platform, with the smart camera and range sensor both pointing downwards. Their models and characteristics are specified in the following sections. The platform or body frame is defined at the IMU frame, and the other sensor frames are calibrated off-line with respect to it. All the image processing to obtain the flow is done by the smart camera.

We aim at tracking the vehicle's state by integrating IMU measurements, and to correct these estimates with velocity (i.e., either 2D planar velocities or image flow) and range readings, observing in turn the IMU biases for their proper compensation. We then use the estimated state to perform closed-loop control of the aerial platform. An overview of the architecture is depicted in Fig. 2.1.

The overall estimation system acts as an odometer that provides absolute altitude, velocity, orientation, angular rate and acceleration, with respect to a precise gravity reference, at a rate of 100 Hz. The x and y positions and the *yaw* angle are not

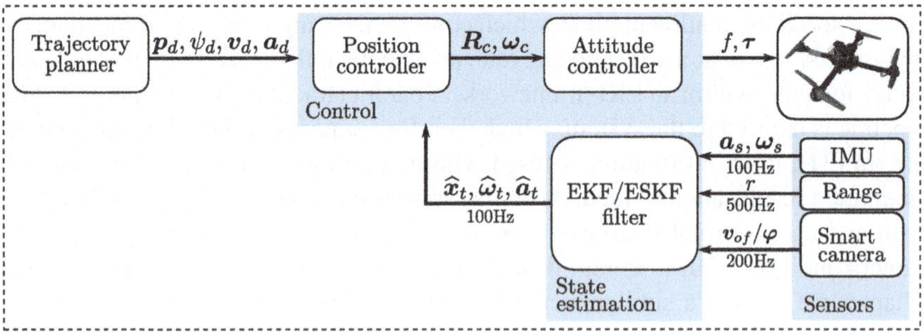

Fig. 2.1 Overview of the architecture pipeline for state estimation and control of the aerial vehicle

observable, and their output is the result of an incremental estimation subject to drift —these modes can be observed with a lower update rate by a higher level task, such as a visual servoing [36], explained in the following chapters.

When compared to other visual-inertial approaches, the optical-flow nature of the smart camera, with a very narrow field of view (FoV) of only 64×64 pixels or 1.6° (compared to the 90° typical of VIO), represents an important limitation, in the sense that visual features are only exploited locally both in space and time (i.e., there is no notion of multiple features being tracked over long periods of time). The number and length of the feature tracks are key to the high precision attained by the more sophisticated VIO methods, and in consequence, we cannot expect equivalent performances. By contrast, the number and length of these feature tracks are the ones responsible for the algorithmic complexity and CPU consumption. In our case, the high filter update rate, made possible by the smart camera and range sensor, and by our light algorithm, contributes to decrease the sensitivity to linearization errors, reducing the accumulation of drift and thus enabling much simpler methods for an acceptable performance.

In this scenario, one key objective of this Chapter is to show that, given a sensor setup with such capabilities, we are able to derive motion estimates that are useful in the mid term (a few minutes, i.e., the typical flight times of a full battery recharge) to drive autonomously the vehicle, without the need to implement complex algorithms. In order to defend the simplest estimation solution, we benchmark a large number of algorithm improvements and variations described in the literature, and show that their impact on system performance is minimal. This should not be surprising given the high frequency of the measurements, but we believe that our benchmarking provides valuable results for establishing good estimation practices.

The algorithm variations that we investigate are shown in Table 2.1, and are properly defined later in the text. They are summarized hereafter, together with the key works that defended them. First, we implement error-state (ESKF) and extended (EKF) Kalman filters [23]. Second, we express the orientation errors of ESKF both in local (LE) and global (GE) frames [16]. Notice how in EKF the orientation error is additive and this distinction is irrelevant. Third, we compare different schemes for rotational motion integration of the quaternion [47], including forward zeroth-order (Q0F), backward zeroth-order (Q0B), and first-order (Q1). Fourth, we compare different integration forms and approximations of the system's transition matrix ($\boldsymbol{F}_1, \boldsymbol{F}_2, \boldsymbol{F}_3$) [17].

Table 2.1 Kalman filters and algorithm variations

Filter type	Quat. error	Quat. integration	Trans. Mat. Trunc.
ESKF	GE, LE	Q0F, Q0B, Q1	$\boldsymbol{F}_1, \boldsymbol{F}_2, \boldsymbol{F}_3$
EKF	–	Q0F, Q0B, Q1	$\boldsymbol{F}_1, \boldsymbol{F}_2, \boldsymbol{F}_3$

In order to describe the state estimation formulation for the ESKF and the EKF in subsequent sections, we present here the following definitions.

In ESKF formulations, we speak of true-, nominal- and error-state, where the error-state values represent the discrepancy between the nominal- and the true-state values. We denote true states with a 't' subindex, $\boldsymbol{x}_t$; nominals with the plain symbol, $\boldsymbol{x}$; and errors with a 'δ' prefix, $\delta\boldsymbol{x}$. These are defined respectively as,

$$\boldsymbol{x}_t = \begin{bmatrix}\boldsymbol{p}_t \ \boldsymbol{v}_t \ \boldsymbol{q}_t \ \boldsymbol{a}_{bt} \ \boldsymbol{\omega}_{bt}\end{bmatrix}^\top \in \mathbb{R}^{16}, \tag{2.1a}$$

$$\boldsymbol{x} = \begin{bmatrix}\boldsymbol{p} \ \boldsymbol{v} \ \boldsymbol{q} \ \boldsymbol{a}_b \ \boldsymbol{\omega}_b\end{bmatrix}^\top \in \mathbb{R}^{16}, \tag{2.1b}$$

$$\delta\boldsymbol{x} = \begin{bmatrix}\delta\boldsymbol{p} \ \delta\boldsymbol{v} \ \delta\boldsymbol{\theta} \ \delta\boldsymbol{a}_b \ \delta\boldsymbol{\omega}_b\end{bmatrix}^\top \in \mathbb{R}^{15}, \tag{2.1c}$$

where $\boldsymbol{p}$, $\boldsymbol{v}$, $\boldsymbol{q}$ are position, velocity and quaternion orientation (refer to Sect. 2.3.1 for quaternion conventions), all expressed in the global world (inertial) frame, $\boldsymbol{\theta}$ are Euler angles, and $\boldsymbol{a}_b$ and $\boldsymbol{\omega}_b$ are accelerometer and gyrometer biases respectively. The error-state is modeled as a Gaussian distribution $\delta\boldsymbol{x} \sim \mathcal{N}(\overline{\delta\boldsymbol{x}}, \boldsymbol{P})$. In the following chapters the notation $\overline{a}$ is adopted to express a variable mean and similarly $\widehat{a}$ its estimated value.

These states are related by the composition

$$\boldsymbol{x}_t = \boldsymbol{x} \oplus \delta\boldsymbol{x}, \tag{2.2}$$

where $\oplus$ wraps the error $\delta\boldsymbol{x}$ onto the manifold of $\boldsymbol{x}$. It corresponds to the trivial addition for all state variables (e.g., $\boldsymbol{p}_t = \boldsymbol{p} + \delta\boldsymbol{p}$) except for the orientation. We use a minimal orientation error $\delta\boldsymbol{\theta} \in \mathfrak{so}(3) \subset \mathbb{R}^3$ living in the space tangent to the special orthogonal group $\mathcal{SO}(3)$ manifold (i.e., in its Lie algebra $\mathfrak{so}(3)$). We contemplate orientation error definitions in the global frame (GE) or in the local frame (LE); their composition is computed respectively with a product *on the left or right hand sides* of the nominal quaternion,

$$\text{global error (GE): } \boldsymbol{q}_t = \delta\boldsymbol{q} \otimes \boldsymbol{q}, \tag{2.3a}$$

$$\text{local error (LE): } \boldsymbol{q}_t = \boldsymbol{q} \otimes \delta\boldsymbol{q}, \tag{2.3b}$$

where $\delta\boldsymbol{q} = \boldsymbol{q}(\delta\boldsymbol{\theta}) \triangleq \exp(\delta\boldsymbol{\theta}/2)$ is the orientation error in $\mathcal{SO}(3)$ expressed as a unit quaternion—see Sect. 2.3.1 for details.

In EKF formulations, we directly estimate the true-state, which is modeled as a Gaussian distribution $\boldsymbol{x}_t \sim \mathcal{N}(\overline{\boldsymbol{x}}_t, \boldsymbol{P})$.

2.3.1 Quaternion Conventions and Properties

We use, as in [42], the Hamilton convention for quaternions, which resumes to defining the three imaginary numbers i, j and k so that $ijk = -1$. If we denote a quaternion ${}^G\boldsymbol{q}_L$ representing the orientation of a local frame L with respect to a global frame G, then a generic composition of two quaternions is defined as

$$\begin{aligned} {}^G\boldsymbol{q}_C &= {}^G\boldsymbol{q}_L \otimes {}^L\boldsymbol{q}_C & \text{(2.4a)} \\ &= {}^G\boldsymbol{Q}_L^+ \, {}^L\boldsymbol{q}_C & \text{(2.4b)} \\ &= {}^L\boldsymbol{Q}_C^- \, {}^G\boldsymbol{q}_L, & \text{(2.4c)} \end{aligned}$$

where, for a quaternion $\boldsymbol{q} = \begin{bmatrix} q_w \; q_x \; q_y \; q_z \end{bmatrix}^\top$, we define $\boldsymbol{Q}^+$ and $\boldsymbol{Q}^-$ respectively as the left- and right- quaternion product matrices,

$$\boldsymbol{Q}^+ = \begin{bmatrix} q_w & -q_x & -q_y & -q_z \\ q_x & q_w & -q_z & q_y \\ q_y & q_z & q_w & -q_x \\ q_z & -q_y & q_x & q_w \end{bmatrix}, \quad \boldsymbol{Q}^- = \begin{bmatrix} q_w & -q_x & -q_y & -q_z \\ q_x & q_w & q_z & -q_y \\ q_y & -q_z & q_w & q_x \\ q_z & q_y & -q_x & q_w \end{bmatrix}. \qquad \text{(2.5a)}$$

Note that, in the quaternion product (2.4a), the right-hand quaternion is defined locally in the frame L, which is specified by the left-hand quaternion.

A vector transformation from a local frame L to the global G is performed by the double product

$$\begin{aligned} {}^G\boldsymbol{\rho} &= {}^G\boldsymbol{q}_L \otimes {}^L\boldsymbol{\rho} \otimes ({}^G\boldsymbol{q}_L)^* & \text{(2.6a)} \\ &= {}^G\boldsymbol{q}_L \otimes {}^L\boldsymbol{\rho} \otimes {}^L\boldsymbol{q}_G\,, & \text{(2.6b)} \end{aligned}$$

where we use the shortcut $\boldsymbol{q} \otimes \boldsymbol{\rho} \equiv \boldsymbol{q} \otimes [0, \boldsymbol{\rho}]^\top$ for convenience of notation, and the notation $(\cdot)^*$ to indicate the conjugate of a quaternion.

Throughout this document, we note $\boldsymbol{q}(o)$ the quaternion and $\boldsymbol{R}(o)$ the rotation matrix equivalents to a generic orientation o. A rotation $\boldsymbol{\theta} = \theta\boldsymbol{u}$, of θ radians around the unit axis $\boldsymbol{u}$, can be expressed in quaternion and matrix forms using the exponential maps

$$\begin{aligned} \boldsymbol{q}(\boldsymbol{\theta}) &= e^{\boldsymbol{\theta}/2} & \text{(2.7a)} \\ &= \begin{bmatrix} \cos(\theta/2) \\ \boldsymbol{u}\sin(\theta/2) \end{bmatrix} & \text{(2.7b)} \\ &\xrightarrow[\theta\to 0]{} \begin{bmatrix} 1 \\ \boldsymbol{\theta}/2 \end{bmatrix}, & \text{(2.7c)} \\ \boldsymbol{R}(\boldsymbol{\theta}) &= e^{[\boldsymbol{\theta}]_\times} & \text{(2.7d)} \\ &= \boldsymbol{I} + \sin\theta[\boldsymbol{u}]_\times + (1-\cos\theta)[\boldsymbol{u}]_\times^2 & \text{(2.7e)} \\ &\xrightarrow[\theta\to 0]{} \boldsymbol{I} + [\boldsymbol{\theta}]_\times\,, & \text{(2.7f)} \end{aligned}$$

with $\boldsymbol{I}$ the identity matrix, and the notation $[\cdot]_\times$ representing the skew-symmetric matrix

$$[\mathbf{a}]_\times = \begin{bmatrix} 0 & -a_z & a_y \\ a_z & 0 & -a_x \\ -a_y & a_z & 0 \end{bmatrix}. \tag{2.8}$$

We also write $\boldsymbol{R} = \boldsymbol{R}(\boldsymbol{q})$, according to

$$\boldsymbol{R}(\boldsymbol{q}) = \begin{bmatrix} q_w^2+q_x^2-q_y^2-q_z^2 & 2(q_xq_y - q_wq_z) & 2(q_xq_z + q_wq_y) \\ 2(q_xq_y + q_wq_z) & q_w^2-q_x^2+q_y^2-q_z^2 & 2(q_yq_z - q_wq_x) \\ 2(q_xq_z - q_wq_y) & 2(q_yq_z + q_wq_x) & q_w^2-q_x^2-q_y^2+q_z^2 \end{bmatrix} \tag{2.9}$$

Finally, the time-derivative of the quaternion is

$$\dot{\boldsymbol{q}} = \frac{1}{2}\boldsymbol{\Omega}(\boldsymbol{\omega})\boldsymbol{q} \tag{2.10}$$

$$= \frac{1}{2}\boldsymbol{q} \otimes \boldsymbol{\omega}, \tag{2.11}$$

with $\boldsymbol{\omega}$ the angular velocity in body frame, and $\boldsymbol{\Omega}$ a skew-symmetric matrix defined as

$$\boldsymbol{\Omega}(\boldsymbol{\omega}) \triangleq \boldsymbol{Q}^-(\boldsymbol{\omega}) = \begin{bmatrix} 0 & -\boldsymbol{\omega}^\top \\ \boldsymbol{\omega} & -[\boldsymbol{\omega}]_\times \end{bmatrix}. \tag{2.12}$$

2.4 Observation Models

The observation models described hereafter are used in the filter propagation (IMU) and correction (smart camera and range sensor) steps.

2.4.1 *Inertial Measurement Unit (IMU)*

The IMU is composed of a 3-axis accelerometer and a 3-axis gyrometer providing, in body frame, acceleration and angular velocity measurements respectively. The accelerometer measures acceleration $\boldsymbol{a}_t$ and gravity $\boldsymbol{g}$ (considered known and constant) together. The gyrometer measures angular rates $\boldsymbol{\omega}_t$, which are already in body frame. These measurements are affected by additive biases, $\boldsymbol{a}_{bt}$ and $\boldsymbol{\omega}_{bt}$, and noises $\boldsymbol{a}_n$ and $\boldsymbol{\omega}_n$. The IMU model reads,

$$\boldsymbol{a}_s = \boldsymbol{R}_t^\top(\boldsymbol{a}_t - \boldsymbol{g}) + \boldsymbol{a}_{bt} + \boldsymbol{a}_n \qquad \in \mathbb{R}^3, \tag{2.13a}$$

$$\boldsymbol{\omega}_s = \boldsymbol{\omega}_t + \boldsymbol{\omega}_{bt} + \boldsymbol{\omega}_n \qquad \in \mathbb{R}^3, \tag{2.13b}$$

where $\boldsymbol{R}_t \triangleq \boldsymbol{R}(\boldsymbol{q}_t)$ is the rotation matrix equivalent to the quaternion $\boldsymbol{q}_t$. The noises are modeled by Gaussian distributions $\boldsymbol{a}_n \sim \mathcal{N}(\mathbf{0}, \boldsymbol{A}_n)$ and $\boldsymbol{\omega}_n \sim \mathcal{N}(\mathbf{0}, \boldsymbol{\Omega}_n)$.

2.4.2 Smart Camera: Sonar Range with 2D Linear Velocity

The smart camera (PX4-Flow) integrates a monocular camera, an ultrasound range sensor aligned with the optical axis, and a triaxial gyrometer. It provides raw optical flow (from now on referred to as just *flow*) at the principal point, a sonar range to a reflective surface, and 3-axial angular rates (for the angular rates, in this Chapter we use the quadrotor's IMU gyrometers because they are of higher quality and already aligned with the vehicle frame). In addition, the smart camera delivers horizontal velocities in metric scale, computed from all the input data. In order to perform correctly, scene lighting must be adequate and the ground is required to have good visual texture.

In a first sensor set we use the IMU together with sonar range readings and 2D linear velocities from the smart camera. With the IMU model shown in Sect. 2.4.1, the observation model of the smart camera for these measurements can be expressed in compact form as

$$h_v(\boldsymbol{x}_t, \boldsymbol{n}_v) = \boldsymbol{S} \begin{bmatrix} \boldsymbol{R}_r^\top \boldsymbol{p}_t + \boldsymbol{p}_r \\ \boldsymbol{R}_c^\top (\boldsymbol{R}_t^\top \boldsymbol{v}_t + [\boldsymbol{\omega}_t]_\times \boldsymbol{p}_c) \end{bmatrix} + \boldsymbol{n}_v , \tag{2.14}$$

where $\boldsymbol{R}_t^\top = \boldsymbol{R}(\boldsymbol{q}_t)^\top$ and $\boldsymbol{n}_v \sim \mathcal{N}(\boldsymbol{0}, \boldsymbol{N}_v)$ is the measurement noise assumed to be Gaussian. The true angular velocity $\boldsymbol{\omega}_t$ (IMU frame) is obtained using (2.13b). $\boldsymbol{p}_r$, $\boldsymbol{p}_c$, $\boldsymbol{R}_r$ and $\boldsymbol{R}_c$ are the calibrated position and orientation of the sonar and camera respectively, calibrated off-line and all expressed in IMU frame. As both sonar range and 2D linear velocities are computed by the same sensor, we can consider $\boldsymbol{p}_r = \boldsymbol{p}_c$ and $\boldsymbol{R}_r = \boldsymbol{R}_t$, but they are here detailed for the sake of completeness. $\boldsymbol{S}$ is a 3×6 matrix to select the required rows defined as

$$\boldsymbol{S} = \begin{bmatrix} \boldsymbol{0}_{3\times2} \; \boldsymbol{I}_{3\times3} \; \boldsymbol{0}_{3\times1} \end{bmatrix} . \tag{2.15}$$

The 2D planar velocity measurement, namely $\boldsymbol{v}_{of}$, is computed directly by the PX4 from its own optical flow $\boldsymbol{\varphi}$, angular rates $\boldsymbol{\omega}_{of}$, and altitude h_{of} measurements, according to

$$\boldsymbol{v}_{of} = \frac{-\boldsymbol{\varphi} \cdot h_{of}}{\Delta t \cdot f} + \boldsymbol{S}' \left(\boldsymbol{\omega}_{of} \times \begin{bmatrix} 0 \; 0 \; h_{of} \end{bmatrix}^\top \right) , \tag{2.16}$$

where f is the focal length of the PX4's camera, expressed in pixels, and $\boldsymbol{S}' = [\boldsymbol{I}_{2x2} \;\; \boldsymbol{0}_{2x1}]$ is also a selection matrix.

Although using the sonar range and the internally computed linear velocities allows us to present a first simple and usable approach, we also want to explore the use of 2D raw flow and add an external infrared (IR) range sensor, for the following reasons:

- The use of internally computed metric velocities has some implications hard to model in the filter because the platform altitude and angular velocities play a role in their computation and some cross correlations arise. In contrast to these linear

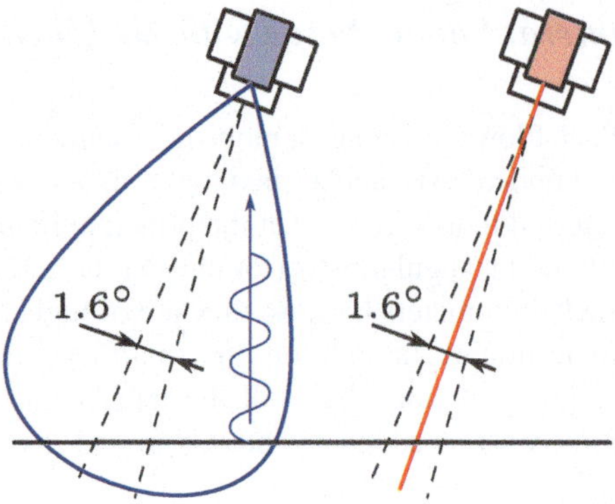

Fig. 2.2 Field of view (FoV) and echo directions using sonar (left) and IR (right) rangers, compared to the camera FoV of 1.6° (dashed). Due to the coarse directivity of the sonar, the strongest echoes are likely to come from a direction perpendicular to the terrain surface, which may fall out of the FoV of the camera, creating false visual-range associations on tilt maneuvers. This is avoided with the excellent directivity of the IR ranger

velocities, the raw flow is not dependent on the filter state thus the process and measurement noises are decoupled.

- The sonar measurements have shown to contain important outliers [35, 38] and we replace the sonar by an IR time-of-flight ranger [35], which has higher precision and directivity. Compared to a sonar implementation (Fig. 2.2-left), the IR ranger (Fig. 2.2-right) has less outliers and ensures that the range echo corresponds to the region observed by the camera (1.6° FoV), deriving in a more predictable behavior that better fits the observation model.

The observation models of both 2D raw flow and IR range sensors are provided in the following.

2.4.3 *Smart Camera: 2D Flow*

Here we take advantage of the raw flow to define an observation function directly in the flow space. The observation model of the flow and the required background are described in the following.

Let w, i and c denote respectively the world, IMU, and camera frames. Let ${}^{c}\boldsymbol{p} = \left[{}^{c}x \; {}^{c}y \; {}^{c}z\right]^{\top}$ be a static 3D point in the ground (${}^{w}z = 0$), expressed in the camera reference frame c as shown in Fig. 2.3. This point is projected onto the image according to the pin-hole model,

$$\varphi = \boldsymbol{P}_f \, \frac{{}^{c}\boldsymbol{p}}{{}^{c}z}, \tag{2.17}$$

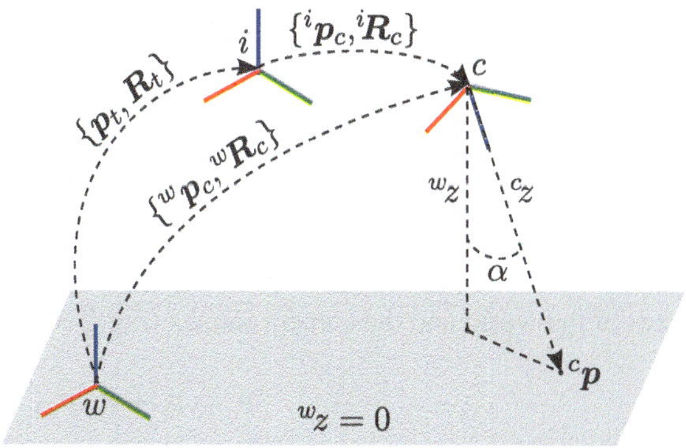

Fig. 2.3 World (w), IMU (i) and camera (c) frames and their relative transformations (curved arrows). The IMU and camera are attached together forming a rigid body, with a twist $\{\boldsymbol{v}_i, \boldsymbol{\omega}_i\}$ at the IMU, which induces a twist $\{\boldsymbol{v}_c, \boldsymbol{\omega}_c\}$ at the camera. A point $\boldsymbol{p}$ on the ground is defined along the camera's optical axis (its Z axis)

with $\boldsymbol{P}_f$ the projection matrix defined in (2.22) and ${}^c z$ the distance, measured along the optical axis, from the optical center to the ground. Taking its time derivative we obtain the optical flow,

$$\boldsymbol{\varphi} = \boldsymbol{P}_f \, \frac{{}^c\dot{\boldsymbol{p}}\,{}^c z - {}^c\boldsymbol{p}\,{}^c\dot{z}}{{}^c z^2} \,. \tag{2.18}$$

The smart camera computes the mean flow in a 64×64 pixels patch centered at the principal point, giving a FoV of $1.6°$ around the optical axis. At the optical axis, we have ${}^c\boldsymbol{p} = \left[0\ 0\ {}^c z\right]^\top$ and (2.18) reduces to

$$\boldsymbol{\varphi} = \boldsymbol{P}_f \, \frac{{}^c\dot{\boldsymbol{p}}}{{}^c z} \,. \tag{2.19}$$

Considering the smart camera in motion with a twist $\{{}^c\boldsymbol{v}_c, {}^c\boldsymbol{\omega}_c\}$ defined in its own frame, the velocity of the point in the camera frame is

$${}^c\dot{\boldsymbol{p}} = -{}^c\boldsymbol{v}_c - {}^c\boldsymbol{\omega}_c \times {}^c\boldsymbol{p} \,. \tag{2.20}$$

Injecting this in (2.19) leads, after minor rearrangements, to

$$\boldsymbol{\varphi} = -\boldsymbol{P}_f \, \frac{{}^c\boldsymbol{v}_c}{{}^c z} + \boldsymbol{P}_\times \, {}^c\boldsymbol{\omega}_c \,, \tag{2.21}$$

where the matrices $\boldsymbol{P}_f$ and $\boldsymbol{P}_\times$ can be expressed in terms of the camera's focal distances f_x, and f_y, measured in horizontal and vertical pixels, respectively, with

$$\boldsymbol{P}_f = \begin{bmatrix} f_x & 0 & 0 \\ 0 & f_y & 0 \end{bmatrix} , \qquad \boldsymbol{P}_\times = \begin{bmatrix} 0 & f_x & 0 \\ -f_y & 0 & 0 \end{bmatrix} . \tag{2.22}$$

To obtain the observation model in terms of our system variables, we can expand ${}^c\boldsymbol{v}_c$, ${}^c\boldsymbol{\omega}_c$ and ${}^c z$ with

$$
{}^{c}\boldsymbol{v}_c = {}^{i}\boldsymbol{R}_c^{\top}\,(\boldsymbol{R}_t^{\top}\,\boldsymbol{v}_t + \boldsymbol{\omega}_t \times {}^{i}\boldsymbol{p}_c)\,, \tag{2.23a}
$$

$$
{}^{c}\boldsymbol{\omega}_c = {}^{i}\boldsymbol{R}_c^{\top}\,\boldsymbol{\omega}_t\,, \tag{2.23b}
$$

$$
{}^{c}z = \frac{{}^{w}z_c}{\cos\alpha} = \frac{{}^{w}\boldsymbol{p}_c(3)}{{}^{w}\boldsymbol{R}_c(3,3)}\,, \tag{2.23c}
$$

where the true angular velocity $\boldsymbol{\omega}_t$ is obtained using (2.13b), the pair $\{{}^{i}\boldsymbol{p}_c, {}^{i}\boldsymbol{R}_c\}$ is the calibrated camera pose expressed in the IMU frame, α is the angle between the Z axes of the world and the camera frames (Fig. 2.3), and

$$
{}^{w}\boldsymbol{p}_c = \boldsymbol{p}_t + \boldsymbol{R}_t\,{}^{i}\boldsymbol{p}_c\,, \tag{2.24a}
$$

$$
{}^{w}\boldsymbol{R}_c = \boldsymbol{R}_t\,{}^{i}\boldsymbol{R}_c\,. \tag{2.24b}
$$

The final observation model considers noisy flow measurements, defined as

$$
h_{\varphi}(\boldsymbol{x}_t, \boldsymbol{n}_{\varphi}) = \varphi(\boldsymbol{x}_t) + \boldsymbol{n}_{\varphi} \quad \in \mathbb{R}^2\,, \tag{2.25}
$$

with a noise $\boldsymbol{n}_{\varphi} \sim \mathcal{N}(\boldsymbol{0}, \boldsymbol{N}_{\varphi})$ assumed to be zero mean Gaussian.

2.4.4 Infrared (IR) Range Sensor

The infrared time-of-flight sensor provides the distance to the surface that bounces its signal (Fig. 2.2-right). When mounted under a quadrotor, facing down and assuming there is no object below the platform, the sensor model is similar to the range presented in (2.23c), but using the ranger frame instead of the camera frame,

$$
h_r(\boldsymbol{x}_t, n_r) = \frac{{}^{w}\boldsymbol{p}_r(3)}{{}^{w}\boldsymbol{R}_r(3,3)} + n_r \quad \in \mathbb{R}\,, \tag{2.26}
$$

with

$$
{}^{w}\boldsymbol{p}_r = \boldsymbol{p}_t + \boldsymbol{R}_t\,{}^{i}\boldsymbol{p}_r\,, \tag{2.27a}
$$

$$
{}^{w}\boldsymbol{R}_r = \boldsymbol{R}_t\,{}^{i}\boldsymbol{R}_r\,, \tag{2.27b}
$$

with noise $n_r \sim \mathcal{N}(0, N_r)$, and where the pair $\{{}^{i}\boldsymbol{p}_r, {}^{i}\boldsymbol{R}_r\}$ is the calibrated pose of the range sensor with respect to the IMU frame. Notice the model difference with respect to the first row of (2.14) which directly retrieves the platform height.

2.5 System Kinematics

As it is common practice in the literature of IMU navigation, e.g.[34], we can define the continuous system kinematic equations $\dot{\boldsymbol{x}}_t = f(\boldsymbol{x}_t, \boldsymbol{u}, \boldsymbol{w})$ as

$$\dot{\boldsymbol{p}}_t = \boldsymbol{v}_t\,, \tag{2.28a}$$
$$\dot{\boldsymbol{v}}_t = \boldsymbol{R}_t\,(\boldsymbol{a}_s - \boldsymbol{a}_{bt} - \boldsymbol{a}_n) + \boldsymbol{g}\,, \tag{2.28b}$$
$$\dot{\boldsymbol{q}}_t = \frac{1}{2}\boldsymbol{q}_t \otimes (\boldsymbol{\omega}_s - \boldsymbol{\omega}_{bt} - \boldsymbol{\omega}_n)\,, \tag{2.28c}$$
$$\dot{\boldsymbol{a}}_{bt} = \boldsymbol{a}_w\,, \tag{2.28d}$$
$$\dot{\boldsymbol{\omega}}_{bt} = \boldsymbol{\omega}_w\,. \tag{2.28e}$$

where we use the shortcut $\boldsymbol{q} \otimes \boldsymbol{\omega} \equiv \boldsymbol{q} \otimes [0, \boldsymbol{\omega}]^\top$ for convenience of notation. This system involves the true-state $\boldsymbol{x}_t$ from (2.1a), is governed by IMU noisy readings $\boldsymbol{u}$ (2.13), and is perturbed by Gaussian noise $\boldsymbol{w}$, the two defined as

$$\boldsymbol{u} = \left[\boldsymbol{a}_s\ \boldsymbol{\omega}_s\right]^\top\,, \tag{2.29a}$$
$$\boldsymbol{w} = \left[\boldsymbol{a}_w\ \boldsymbol{\omega}_w\right]^\top\,, \tag{2.29b}$$

with $\boldsymbol{a}_w \sim \mathcal{N}(\boldsymbol{0}, \boldsymbol{A}_w)$ and $\boldsymbol{\omega}_w \sim \mathcal{N}(\boldsymbol{0}, \boldsymbol{\Omega}_w)$.

In the ESKF, we distinguish between nominal- and error-state kinematics. The nominal kinematics correspond to the modeled system without noises or perturbations

$$\dot{\boldsymbol{p}} = \boldsymbol{v}\,, \tag{2.30a}$$
$$\dot{\boldsymbol{v}} = \boldsymbol{R}\,(\boldsymbol{a}_s - \boldsymbol{a}_b) + \boldsymbol{g}\,, \tag{2.30b}$$
$$\dot{\boldsymbol{q}} = \frac{1}{2}\boldsymbol{q} \otimes (\boldsymbol{\omega}_s - \boldsymbol{\omega}_b)\,, \tag{2.30c}$$
$$\dot{\boldsymbol{a}}_b = \boldsymbol{0}\,, \tag{2.30d}$$
$$\dot{\boldsymbol{\omega}}_b = \boldsymbol{0}\,. \tag{2.30e}$$

To obtain the error kinematics we write each true-state equation in (2.28) as its composition of nominal- and error-states in (2.2) and (2.3), solve for the error-state, and simplify all second-order infinitesimals. For the sake of clarity, the details of this process are presented in Appendix 2.A but are here summarized with the following. The result for the linear velocity and orientation elements ($\delta\boldsymbol{v}$ and $\delta\boldsymbol{\theta}$) depend on the orientation error representation (GE or LE). With GE we have

$$\delta\dot{\boldsymbol{p}} = \delta\boldsymbol{v}\,, \tag{2.31a}$$
$$\delta\dot{\boldsymbol{v}} = -[\boldsymbol{R}(\boldsymbol{a}_s - \boldsymbol{a}_b)]_\times\,\delta\boldsymbol{\theta} - \boldsymbol{R}\,\delta\boldsymbol{a}_b - \boldsymbol{R}\boldsymbol{a}_n\,, \tag{2.31b}$$
$$\delta\dot{\boldsymbol{\theta}} = -\boldsymbol{R}\delta\boldsymbol{\omega}_b - \boldsymbol{R}\,\boldsymbol{\omega}_n\,, \tag{2.31c}$$

$$\delta\dot{a}_b = a_w\,, \tag{2.31d}$$
$$\delta\dot{\omega}_b = \omega_w\,, \tag{2.31e}$$

whereas with LE we need to replace (2.31b) and (2.31c) above with

$$\delta\dot{v} = -R\,[a_s - a_b]_\times\,\delta\theta - R\,\delta a_b - Ra_n\,, \tag{2.31f}$$
$$\delta\dot{\theta} = -[\omega_s - \omega_b]_\times\,\delta\theta - \delta\omega_b - \omega_n\,. \tag{2.31g}$$

2.5.1 System Kinematics in Discrete Time

One of the aims of this Chapter is to analyze the impact of using different integration approximations of the previous equations. Integrating continuous differential equations of the type $\dot{x} = f(x, u)$ from time $(k-1)\Delta t$ to $k\Delta t$ can be done in a number of ways. A common technique is to integrate the linearized system, $\dot{x} = Ax + Bu$, with $A = \partial f/\partial x$, $B = \partial f/\partial u$, into the discrete-time $x_k \approx F x_{k-1} + B u \Delta t$, with $F = e^{A\Delta t}$, and to truncate the exponential Taylor series $e^{A\Delta t} = \sum A^n \Delta t^n/n!$ at different orders, obtaining the different approximations F_N of the filter transition matrix,

$$F_N \triangleq \sum_{n=0}^{N<\infty} \frac{1}{n!} A^n \Delta t^n = I + A\Delta t + \frac{1}{2} A^2 \Delta t^2 + \ldots + \frac{1}{N!} A^N \Delta t^N\,. \tag{2.32}$$

We provide extensive details in Sect. 2.5.2.

For the quaternion, it is possible and convenient, through the exponential maps, to obtain closed-form expressions of the infinite Taylor series [42, 47]. For this reason, we contemplate here the *zero-th forward*, *zero-th backward* and *first* order integrators of (2.28c), that we name Q0F, Q0B and Q1 respectively,

$$\text{Q0F: } q_k \approx q_{k-1} \otimes q(\omega_{k-1}\Delta t)\,, \tag{2.33a}$$
$$\text{Q0B: } q_k \approx q_{k-1} \otimes q(\omega_k \Delta t)\,, \tag{2.33b}$$
$$\text{Q1: } q_k \approx q_{k-1} \otimes \left(q(\bar{\omega}\Delta t) + \frac{\Delta t^2}{24} \begin{bmatrix} 0 \\ \omega_{k-1} \times \omega_k \end{bmatrix} \right), \tag{2.33c}$$

with $\omega_k \triangleq \omega_{s,k} - \omega_{bt,k}$, $\omega_{bt,k} \approx \omega_{bt,k-1}$ and $\bar{\omega} \triangleq (\omega_{k-1} + \omega_k)/2$, and $q(\cdot)$ as defined in (2.7c). Notice that Q0F and Q0B are proper $\mathcal{SO}(3)$ integrators: they integrate the angular rates in $\mathfrak{so}(3)$ using forward or backward Euler, producing an angular step $\Delta\theta = \omega\Delta t \in \mathfrak{so}(3)$, and construct the quaternion step $\Delta q = q(\Delta\theta) = \exp(\Delta\theta/2)$ using the exponential map (2.7c), which is then composed in the $\mathcal{SO}(3)$ group locally (i.e., at the right side of the product). Q1 accounts for second-order terms appearing only when the rotation axis changes direction within the integration interval (i.e., $\omega_{k-1} \times \omega_k \neq 0$).

Integrations for the ESKF and the EKF are detailed in the following paragraphs. For the sake of clarity, in this section we integrate the kinematic equations using only $\boldsymbol{F}_1$ (also known as backward Euler integration) for all variables except the quaternion. For the quaternion, we use Q0B. This choice is pertinent: as will be revealed in the benchmarking, improving the approximations of the transition matrix and the quaternion beyond the forms presented in this section has only minimal effect on the overall performance.

2.5.1.1 Error-State Kalman Filter Kinematics

For the ESKF we need to integrate the nominal- and the error-state equations. The integration of the nominal-state equations (2.30) results in

$$\boldsymbol{p} \leftarrow \boldsymbol{p} + \boldsymbol{v}\,\Delta t\,, \tag{2.34a}$$
$$\boldsymbol{v} \leftarrow \boldsymbol{v} + (\boldsymbol{R}\,(\boldsymbol{a}_s - \boldsymbol{a}_b) + \boldsymbol{g})\,\Delta t\,, \tag{2.34b}$$
$$\boldsymbol{q} \leftarrow \boldsymbol{q} \otimes \boldsymbol{q}((\boldsymbol{\omega}_s - \boldsymbol{\omega}_b)\Delta t)\,, \tag{2.34c}$$
$$\boldsymbol{a}_b \leftarrow \boldsymbol{a}_b\,, \tag{2.34d}$$
$$\boldsymbol{\omega}_b \leftarrow \boldsymbol{\omega}_b\,, \tag{2.34e}$$

where "$\leftarrow$" stands for "gets updated with", i.e. $x \leftarrow f(x, \cdot)$ is equivalent to $x_k = f(x_{k-1}, \cdot_k)$. Similarly, the integration of the error-state equations (2.31) produces, for a globally-defined error (GE),

$$\delta\boldsymbol{p} \leftarrow \delta\boldsymbol{p} + \delta\boldsymbol{v}\,\Delta t\,, \tag{2.35a}$$
$$\delta\boldsymbol{v} \leftarrow \delta\boldsymbol{v} - ([\boldsymbol{R}\,(\boldsymbol{a}_s - \boldsymbol{a}_b)]_\times \delta\boldsymbol{\theta} + \boldsymbol{R}\delta\boldsymbol{a}_b)\,\Delta t + \boldsymbol{v}_i\,, \tag{2.35b}$$
$$\delta\boldsymbol{\theta} \leftarrow \delta\boldsymbol{\theta} - \boldsymbol{R}\delta\boldsymbol{\omega}_b\Delta t + \boldsymbol{\theta}_i\,, \tag{2.35c}$$
$$\delta\boldsymbol{a}_b \leftarrow \delta\boldsymbol{a}_b + \boldsymbol{a}_i\,, \tag{2.35d}$$
$$\delta\boldsymbol{\omega}_b \leftarrow \delta\boldsymbol{\omega}_b + \boldsymbol{\omega}_i\,, \tag{2.35e}$$

whereas for LE,

$$\delta\boldsymbol{v} \leftarrow \delta\boldsymbol{v} - (\boldsymbol{R}[(\boldsymbol{a}_s - \boldsymbol{a}_b)]_\times \delta\boldsymbol{\theta} + \boldsymbol{R}\delta\boldsymbol{a}_b)\Delta t + \boldsymbol{v}_i\,, \tag{2.35f}$$
$$\delta\boldsymbol{\theta} \leftarrow \delta\boldsymbol{\theta} - ([\boldsymbol{\omega}_s - \boldsymbol{\omega}_b]_\times\,\delta\boldsymbol{\theta} + \delta\boldsymbol{\omega}_b)\Delta t + \boldsymbol{\theta}_i\,. \tag{2.35g}$$

Here, $\boldsymbol{v}_i, \boldsymbol{\theta}_i, \boldsymbol{a}_i$ and $\boldsymbol{\omega}_i$ are random impulses applied to the velocity, orientation and bias estimates, modeled with Gaussian processes. Their mean is zero, and their covariances matrices are obtained by integrating the variances of the IMU measurement noises, $\boldsymbol{a}_n$, $\boldsymbol{\omega}_n$, and the IMU bias random walks, $\boldsymbol{a}_\omega$, $\boldsymbol{\omega}_\omega$, over the time step Δt,

$$\begin{aligned}
\boldsymbol{V}_i &= \boldsymbol{A}_n\,\Delta t^2 = \sigma_{a_n}^2\,\Delta t^2\,\boldsymbol{I} & [m^2/s^2]\,, && (2.36a)\\
\boldsymbol{\Theta}_i &= \boldsymbol{\Omega}_n\,\Delta t^2 = \sigma_{\omega_n}^2\,\Delta t^2\,\boldsymbol{I} & [rad^2]\,, && (2.36b)\\
\boldsymbol{A}_i &= \boldsymbol{A}_w\,\Delta t = \sigma_{a_w}^2\,\Delta t\,\boldsymbol{I} & [m^2/s^4]\,, && (2.36c)\\
\boldsymbol{\Omega}_i &= \boldsymbol{\Omega}_w\,\Delta t = \sigma_{\omega_w}^2\,\Delta t\,\boldsymbol{I} & [rad^2/s^2]\,, && (2.36d)
\end{aligned}$$

where $\sigma_{a_n}[m/s^2]$, $\sigma_{\omega_n}[rad/s]$, $\sigma_{a_w}[m/s^2\sqrt{s}]$ and $\sigma_{\omega_w}[rad/s\sqrt{s}]$ are to be determined from the information in the IMU datasheet, from real measurements, or –preferably as a last resort– via filter tuning.

2.5.1.2 Extended Kalman Filter Kinematics

In this case, we simply integrate the true-state kinematic equations (2.28). Notice that the result is equivalent to the nominal integration in ESKF, but incorporating the noises $\boldsymbol{v}_i$, $\boldsymbol{\theta}_i$, and random walk biases $\boldsymbol{a}_i$, $\boldsymbol{\omega}_i$,

$$\begin{aligned}
\boldsymbol{p}_t &\leftarrow \boldsymbol{p}_t + \boldsymbol{v}_t\,\Delta t\,, && (2.37a)\\
\boldsymbol{v}_t &\leftarrow \boldsymbol{v}_t + (\boldsymbol{R}_t\,(\boldsymbol{a}_s - \boldsymbol{a}_{bt}) + \boldsymbol{g})\,\Delta t + \boldsymbol{v}_i\,, && (2.37b)\\
\boldsymbol{q}_t &\leftarrow \boldsymbol{q}_t \otimes \boldsymbol{q}\{(\boldsymbol{\omega}_s - \boldsymbol{\omega}_{bt})\Delta t + \boldsymbol{\theta}_i\}\,, && (2.37c)\\
\boldsymbol{a}_{bt} &\leftarrow \boldsymbol{a}_{bt} + \boldsymbol{a}_i\,, && (2.37d)\\
\boldsymbol{\omega}_{bt} &\leftarrow \boldsymbol{\omega}_{bt} + \boldsymbol{\omega}_i\,. && (2.37e)
\end{aligned}$$

2.5.2 *Filter Transition Matrices*

We detail the construction of the filter transition matrix for the three involved integrals: ESKF nominal- (2.34), ESKF error- (2.35), and EKF true- (2.37) kinematics. For each case, we need to define the matrix $\boldsymbol{A}$ as the Jacobian of the respective continuous-time system, and build the transition matrix $\boldsymbol{F}_N$ as the truncated Taylor series (2.32). In the following paragraphs, we detail the matrices $\boldsymbol{A}$ for each case, and some examples of their first powers up to $n = 3$. The reader should find no difficulties in building the powers of $\boldsymbol{A}$ that have not been detailed, and the transition matrices $\boldsymbol{F}_N$ using the Taylor series from (2.32).

2.5.2.1 ESKF Transition Matrix

The Jacobian $\boldsymbol{A} = \partial f(\boldsymbol{x}, \delta\boldsymbol{x}, \cdot)/\partial\delta\boldsymbol{x}$ of the ESKF's continuous time error-state system $f(\cdot)$ (2.31) using GE is,

$$A = \begin{bmatrix} 0 & \boldsymbol{I} & 0 & 0 & 0 \\ 0 & 0 & \boldsymbol{V} & -\boldsymbol{R} & 0 \\ 0 & 0 & 0 & 0 & -\boldsymbol{R} \\ 0 & 0 & 0 & 0 & 0 \\ 0 & 0 & 0 & 0 & 0 \end{bmatrix} , \tag{2.38}$$

with

$$\boldsymbol{V} = -[\boldsymbol{R}(\boldsymbol{a}_s - \boldsymbol{a}_b)]_\times . \tag{2.39}$$

Its powers are,

$$\boldsymbol{A}^2 = \begin{bmatrix} 0 & 0 & \boldsymbol{V} & -\boldsymbol{R} & 0 \\ 0 & 0 & 0 & 0 & -\boldsymbol{V}\boldsymbol{R} \\ 0 & 0 & 0 & 0 & 0 \\ 0 & 0 & 0 & 0 & 0 \\ 0 & 0 & 0 & 0 & 0 \end{bmatrix} , \ \boldsymbol{A}^3 = \begin{bmatrix} 0 & 0 & 0 & 0 & -\boldsymbol{V}\boldsymbol{R} \\ 0 & 0 & 0 & 0 & 0 \\ 0 & 0 & 0 & 0 & 0 \\ 0 & 0 & 0 & 0 & 0 \\ 0 & 0 & 0 & 0 & 0 \end{bmatrix} ,$$

and $\boldsymbol{A}^n = \boldsymbol{0}$ for $n > 3$.

For LE we have

$$\boldsymbol{A} = \begin{bmatrix} 0 & \boldsymbol{I} & 0 & 0 & 0 \\ 0 & 0 & \boldsymbol{V} & -\boldsymbol{R} & 0 \\ 0 & 0 & \boldsymbol{\Theta} & 0 & -\boldsymbol{I} \\ 0 & 0 & 0 & 0 & 0 \\ 0 & 0 & 0 & 0 & 0 \end{bmatrix} , \ \boldsymbol{A}^2 = \begin{bmatrix} 0 & 0 & \boldsymbol{V} & -\boldsymbol{R} & 0 \\ 0 & 0 & \boldsymbol{V}\boldsymbol{\Theta} & 0 & -\boldsymbol{V} \\ 0 & 0 & \boldsymbol{\Theta}^2 & 0 & -\boldsymbol{\Theta} \\ 0 & 0 & 0 & 0 & 0 \\ 0 & 0 & 0 & 0 & 0 \end{bmatrix} , \cdots ,$$

with

$$\boldsymbol{V} = -\boldsymbol{R}[\boldsymbol{a}_s - \boldsymbol{a}_b]_\times , \tag{2.40a}$$

$$\boldsymbol{\Theta} = -[\boldsymbol{\omega}_s - \boldsymbol{\omega}_b]_\times . \tag{2.40b}$$

2.5.2.2 EKF Transition Matrix

The Jacobians $\boldsymbol{A} = \partial f(\boldsymbol{x}, \cdot)/\partial \boldsymbol{x}$ of the continuous-time EKF true- (2.28) and ESKF nominal- (2.30) systems are equal to each other, having

$$\boldsymbol{A} = \begin{bmatrix} 0 & \boldsymbol{I} & 0 & 0 & 0 \\ 0 & 0 & \boldsymbol{V} & -\boldsymbol{R} & 0 \\ 0 & 0 & \boldsymbol{W} & 0 & \boldsymbol{Q} \\ 0 & 0 & 0 & 0 & 0 \\ 0 & 0 & 0 & 0 & 0 \end{bmatrix} , \ \boldsymbol{A}^2 = \begin{bmatrix} 0 & 0 & \boldsymbol{V} & -\boldsymbol{R} & 0 \\ 0 & 0 & \boldsymbol{V}\boldsymbol{W} & 0 & \boldsymbol{V}\boldsymbol{Q} \\ 0 & 0 & \boldsymbol{W}^2 & 0 & \boldsymbol{W}\boldsymbol{Q} \\ 0 & 0 & 0 & 0 & 0 \\ 0 & 0 & 0 & 0 & 0 \end{bmatrix} , \cdots ,$$

where $\boldsymbol{V}$, $\boldsymbol{W}$ and $\boldsymbol{Q}$ are defined by

$$V = \frac{\partial \boldsymbol{R}\{\boldsymbol{q}\}\,(\boldsymbol{a}_s - \boldsymbol{a}_b)}{\partial \boldsymbol{q}}, \tag{2.41a}$$

$$W = \frac{\partial \frac{1}{2}\boldsymbol{q} \otimes (\boldsymbol{\omega}_s - \boldsymbol{\omega}_b)}{\partial \boldsymbol{q}}, \tag{2.41b}$$

$$Q = \frac{\partial \frac{1}{2}\boldsymbol{q} \otimes (\boldsymbol{\omega}_s - \boldsymbol{\omega}_b)}{\partial \boldsymbol{\omega}_b}, \tag{2.41c}$$

and are developed hereafter. For the first Jacobian $\boldsymbol{V}$ it is convenient to recall the derivative of a rotation of a vector $\mathbf{a}$ by a quaternion $\boldsymbol{q} = [w, x, y, z]^\top = [w, \boldsymbol{\rho}]^\top$, with respect to the quaternion,

$$\boldsymbol{V}(\boldsymbol{q}, \boldsymbol{a}) \triangleq \frac{\partial \boldsymbol{R}\{\boldsymbol{q}\}\boldsymbol{a}}{\partial \boldsymbol{q}} = \frac{\partial (\boldsymbol{q} \otimes \boldsymbol{a} \otimes \boldsymbol{q}^*)}{\partial \boldsymbol{q}} \tag{2.42}$$

$$= 2\big[w\boldsymbol{a} + \boldsymbol{\rho} \times \boldsymbol{a} \mid \boldsymbol{\rho}\boldsymbol{a}^\top - \boldsymbol{a}\boldsymbol{\rho}^\top + \boldsymbol{a}^\top\boldsymbol{\rho}\boldsymbol{I}_3 - w[\boldsymbol{a}]_\times\big], \tag{2.43}$$

having therefore

$$\boldsymbol{V} = \boldsymbol{V}(\boldsymbol{q},\, \boldsymbol{a}_s - \boldsymbol{a}_b)\,. \tag{2.44}$$

For the Jacobian $\boldsymbol{W}$ we have from (2.10)

$$\boldsymbol{W} = \frac{1}{2}\boldsymbol{\Omega}(\boldsymbol{\omega}_s - \boldsymbol{\omega}_b)\,, \tag{2.45}$$

with $\boldsymbol{\Omega}(\boldsymbol{\omega})$ the skew-symmetric matrix defined in (2.12). Finally, for the Jacobian $\boldsymbol{Q}$ we use (2.4), (2.5) and (2.7c) to obtain

$$\boldsymbol{Q} = -\frac{1}{2}\begin{bmatrix} -x & -y & -z \\ w & -z & y \\ z & w & -x \\ -y & x & w. \end{bmatrix} \tag{2.46}$$

2.6 Error-State Kalman Filter

We are interested in estimating the true-state $\boldsymbol{x}_t$, which we do as follows. High-frequency IMU data is integrated into the nominal-state $\boldsymbol{x}$, which does not take into account noise terms or other possible model imperfections and, as a consequence, it accumulates errors. These errors are collected in the error-state, defined as the multivariate Gaussian

$$\delta\boldsymbol{x} \sim \mathcal{N}(\overline{\delta\boldsymbol{x}}, \boldsymbol{P})\,, \tag{2.47}$$

this time incorporating all the noise and perturbations. In parallel with integration of the nominal-state, the ESKF predicts a prior estimate of this error-state, and uses the other sensor readings (sonar range and 2D velocity, or flow and IR range) in a correction phase to provide a posterior. After each correction, the error-state's estimate $\widehat{\delta \boldsymbol{x}}$ is injected into the nominal-state, and then reset to zero. Because of this reset, at each time the nominal state $\boldsymbol{x}$ is the best estimate of the true state $\boldsymbol{x}_t$, and the estimated uncertainty is described by the error covariances matrix $\boldsymbol{P}$.

2.6.1 ESKF: Filter Prediction

Apart from the true-, nominal- and error-state vectors, it is convenient here to consider our kinematic models in a generic form $\boldsymbol{x}_t \leftarrow f(\boldsymbol{x}_t, \boldsymbol{u}, \boldsymbol{i})$ that we will identify with the appropriate equation numbers. The input vector $\boldsymbol{u}$ (IMU readings) and the perturbation impulses vector $\boldsymbol{i}$ are defined as follows

$$\boldsymbol{u} = \begin{bmatrix} \boldsymbol{a}_s \\ \boldsymbol{\omega}_s \end{bmatrix}, \quad \boldsymbol{i} = \begin{bmatrix} \boldsymbol{v}_i \\ \boldsymbol{\omega}_i \\ \boldsymbol{a}_i \\ \boldsymbol{\omega}_i \end{bmatrix}. \tag{2.48}$$

At the arrival of a new IMU measurement, we propagate the nominal-state $\boldsymbol{x}$ according to a version of (2.34) using the selected $\boldsymbol{F}_N$,

$$\boldsymbol{x} \leftarrow f(\boldsymbol{x}, \boldsymbol{u}, \boldsymbol{0}), \tag{2.49}$$

and the error-state Gaussian with the filter using (2.35),

$$\widehat{\delta \boldsymbol{x}} \leftarrow \boldsymbol{F}_N \, \widehat{\delta \boldsymbol{x}}, \tag{2.50a}$$

$$\boldsymbol{P} \leftarrow \boldsymbol{F}_N \, \boldsymbol{P} \, \boldsymbol{F}_N^\top + \boldsymbol{F}_i \, \boldsymbol{Q}_i \, \boldsymbol{F}_i^\top, \tag{2.50b}$$

where $\boldsymbol{F}_N$ is the transition matrix (i.e., the Jacobian of (2.35) with respect to the error-state $\delta \boldsymbol{x}$—see Sect. 2.5.2 for details), $\boldsymbol{F}_i$ is the Jacobian of (2.35) with respect to the perturbations vector $\boldsymbol{i}$, obtained by simple inspection, and $\boldsymbol{Q}_i$ is the covariances matrix of the perturbation impulses, given by

$$\boldsymbol{Q}_i = \begin{bmatrix} \boldsymbol{V}_i & \boldsymbol{0} & \boldsymbol{0} & \boldsymbol{0} \\ \boldsymbol{0} & \boldsymbol{\Theta}_i & \boldsymbol{0} & \boldsymbol{0} \\ \boldsymbol{0} & \boldsymbol{0} & \boldsymbol{A}_i & \boldsymbol{0} \\ \boldsymbol{0} & \boldsymbol{0} & \boldsymbol{0} & \boldsymbol{\Omega}_i \end{bmatrix}. \tag{2.51}$$

Notice how the Eq. (2.50a) for $\widehat{\delta \boldsymbol{x}}$ can be neglected, thus the error estimate $\widehat{\delta \boldsymbol{x}}$ starts at zero (right element) due to the reset operation performed in the previous time step, after injecting the error estimate into the nominal state.

2.6.2 ESKF: Filter Innovation and Correction

We consider the arrival of sensor data other than IMU, with a model

$$y = h_j(x_t, n_j)\,, \tag{2.52}$$

with $j = v$ for the sonar and 2D velocity observation model (2.14), $j = \varphi$ for the flow sensor observation model (2.21, 2.25), and $j = r$ for the range sensor model (2.26). Since $\widehat{\delta x} = 0$, we have $\widehat{x}_t = x$, and the innovation z and its covariance Z read

$$z = y - h_j(x, \mathbf{0})\,, \tag{2.53a}$$

$$Z = H_j P H_j^\top + N_j\,, \tag{2.53b}$$

with $H_j = \partial h_j / \partial \delta x$ being the observation Jacobian of sonar range with 2D linear velocity (2.14), flow (2.21) or IR range (2.26), defined with respect to the error-state δx. These derivations follow standard procedures and are here avoided, instead we include in Appendix 2.B.2 some useful partial derivatives to compute them.

In order to be robust to possible measurement outliers, we perform a χ^2-test based on the Mahalanobis distance of the innovation [1]. Inliers are validated by checking the condition

$$z^\top Z^{-1} z \leq \chi^2_{th}\,, \tag{2.54}$$

with χ^2_{th} equal to the 0.95 probability quantile of the χ^2 distribution. If we pass the test, we proceed by computing the Kalman gain K and observing the filter error,

$$K = P H^\top Z^{-1}\,, \tag{2.55a}$$

$$\widehat{\delta x} \leftarrow K z\,, \tag{2.55b}$$

$$P \leftarrow P - K Z K^\top\,. \tag{2.55c}$$

Note how symmetry can be enforced in (2.55c), e.g.using the Joseph form

$$P \leftarrow (I - KH) P (I - KH)^\top + K Z K^\top\,. \tag{2.56}$$

2.6.3 ESKF: Injection of the Observed Error into the Nominal-State

After an ESKF update, we update the nominal-state with the observed error using the appropriate compositions $x \leftarrow x \oplus \widehat{\delta x}$ introduced in (2.2). This operation is a simple addition for most variables except for the orientation, which depends on the error representation. Hence, for a global definition (GE) we have

$$q \leftarrow q(\widehat{\delta\theta}) \otimes q\,, \tag{2.57a}$$

whereas for a LE,

$$q \leftarrow q \otimes q(\widehat{\delta\theta})\,. \tag{2.57b}$$

Notice that these operations constitute proper updates in the $\mathscr{SO}(3)$ manifold represented by unit quaternions.

2.6.4 ESKF: Reset Operation

In ESKF, after injecting the error into the nominal-state, the error-state estimate $\widehat{\delta x}$ gets reset to zero. This is especially relevant for the orientation part, as the new orientation error will be expressed locally with respect to the orientation frame of the new nominal-state. To make the ESKF update complete, the covariance of the error needs to be updated according to this modification.

Let us call the error reset function $g(\cdot)$ written as follows,

$$\delta x \leftarrow g(\delta x) = \delta x \ominus \widehat{\delta x}\,, \tag{2.58}$$

where the generic operation $\ominus$ stands for the inverse composition of $\oplus$. The ESKF reset operation is

$$\widehat{\delta x} \leftarrow \mathbf{0}\,, \tag{2.59a}$$

$$P(\theta) \leftarrow G P(\theta) G^\top\,, \tag{2.59b}$$

where $P(\theta)$ is the orientation block of the covariance matrix P, and G is the Jacobian matrix of $g(\cdot)$ for the orientation term, with respect to the error-state. If the orientation error is defined globally (GE), this Jacobian corresponds to

$$G = I + \left[\frac{1}{2}\widehat{\delta\theta}\right]_\times\,. \tag{2.60}$$

whereas with LE

$$G = I - \left[\frac{1}{2}\widehat{\delta\theta}\right]_\times\,, \tag{2.61}$$

The details of this derivation are shown in the following.

2.6.4.1 ESKF: Reset Jacobian

We want to obtain the expression of the new angular error $\delta\boldsymbol{\theta}^+$ with respect to the old error $\delta\boldsymbol{\theta}$ and the observed error $\widehat{\delta\boldsymbol{\theta}}$. To proceed we have to consider the following:

(a) The true orientation does not change on error reset, i.e. $\boldsymbol{q}_t^+ = \boldsymbol{q}_t$, thus depending on how the orientation error is expressed:

$$GE : \delta\boldsymbol{q}^+ \otimes \boldsymbol{q}^+ = \delta\boldsymbol{q} \otimes \boldsymbol{q}\,, \tag{2.62a}$$
$$LE : \boldsymbol{q}^+ \otimes \delta\boldsymbol{q}^+ = \boldsymbol{q} \otimes \delta\boldsymbol{q}\,. \tag{2.62b}$$

(b) The observed error mean has been injected into the nominal state and depending on the orientation error expression:

$$GE : \boldsymbol{q}^+ = \widehat{\delta\boldsymbol{q}} \otimes \boldsymbol{q}\,, \tag{2.63a}$$
$$LE : \boldsymbol{q}^+ = \boldsymbol{q} \otimes \widehat{\delta\boldsymbol{q}}\,. \tag{2.63b}$$

- **ESKF reset Jacobian with GE**

Combining the previous statements we obtain an expression of the new orientation error as

$$\delta\boldsymbol{q}^+ = \delta\boldsymbol{q} \otimes \widehat{\delta\boldsymbol{q}}^*\,, \tag{2.64a}$$
$$= \boldsymbol{Q}^-(\widehat{\delta\boldsymbol{q}}^*)\delta\boldsymbol{q}\,. \tag{2.64b}$$

Considering $\widehat{\delta\boldsymbol{q}}^* \approx \left[1\ -\frac{1}{2}\widehat{\delta\boldsymbol{\theta}}\right]^\top$, and removing all quadratic terms, the identity above can be expanded as

$$\begin{bmatrix} 1 \\ \frac{1}{2}\delta\boldsymbol{\theta}^+ \end{bmatrix} = \begin{bmatrix} 1 & \frac{1}{2}\widehat{\delta\boldsymbol{\theta}}^\top \\ -\frac{1}{2}\widehat{\delta\boldsymbol{\theta}} & \boldsymbol{I} + \left[\frac{1}{2}\widehat{\delta\boldsymbol{\theta}}\right]_\times \end{bmatrix} \begin{bmatrix} 1 \\ \frac{1}{2}\delta\boldsymbol{\theta} \end{bmatrix}. \tag{2.65}$$

The first row verifies the orthogonality principle (i.e., $\widehat{\delta\boldsymbol{\theta}}^\top \delta\boldsymbol{\theta} = \boldsymbol{0}$), which is a necessary and sufficient condition for the optimality of a Bayesian estimator [1]. From the rest of the elements we can extract

$$\delta\boldsymbol{\theta}^+ = -\widehat{\delta\boldsymbol{\theta}} + \left(\boldsymbol{I} + \left[\frac{1}{2}\widehat{\delta\boldsymbol{\theta}}\right]_\times\right)\delta\boldsymbol{\theta}\,, \tag{2.66}$$

where by simple inspection the Jacobian is

$$\boldsymbol{G} = \boldsymbol{I} + \left[\frac{1}{2}\widehat{\delta\boldsymbol{\theta}}\right]_\times\,. \tag{2.67}$$

- **ESKF reset Jacobian with LE**

Similarly to GE, we can combine the previous statements to obtain an expression for $\delta\boldsymbol{\theta}^+$. Then, we have

$$\begin{aligned}
\delta\boldsymbol{q}^+ &= (\boldsymbol{q}^+)^* \otimes \boldsymbol{q} \otimes \delta\boldsymbol{q} && \text{(2.68a)}\\
&= (\boldsymbol{q} \otimes \widehat{\delta\boldsymbol{q}})^* \otimes \boldsymbol{q} \otimes \delta\boldsymbol{q} && \text{(2.68b)}\\
&= \widehat{\delta\boldsymbol{q}}^* \otimes \delta\boldsymbol{q} && \text{(2.68c)}\\
&= \boldsymbol{Q}^+(\widehat{\delta\boldsymbol{q}}^*)\delta\boldsymbol{q}\,. && \text{(2.68d)}
\end{aligned}$$

Considering $\widehat{\delta\boldsymbol{q}}^* \approx \left[1 \; -\frac{1}{2}\widehat{\delta\boldsymbol{\theta}}\right]^\top$, and removing all quadratic terms, the identity above can be expanded as

$$\begin{bmatrix} 1 \\ \frac{1}{2}\delta\boldsymbol{\theta}^+ \end{bmatrix} = \begin{bmatrix} 1 & \frac{1}{2}\widehat{\delta\boldsymbol{\theta}}^\top \\ -\frac{1}{2}\widehat{\delta\boldsymbol{\theta}} & \boldsymbol{I} - \left[\frac{1}{2}\widehat{\delta\boldsymbol{\theta}}\right]_\times \end{bmatrix} \begin{bmatrix} 1 \\ \frac{1}{2}\delta\boldsymbol{\theta} \end{bmatrix}. \tag{2.69}$$

A for the case of the ESKF reset Jacobian with GE, the first row verifies the orthogonality principle. From the rest of the elements we can extract

$$\delta\boldsymbol{\theta}^+ = -\widehat{\delta\boldsymbol{\theta}} + \left(\boldsymbol{I} - \left[\frac{1}{2}\widehat{\delta\boldsymbol{\theta}}\right]_\times\right)\delta\boldsymbol{\theta}\,, \tag{2.70}$$

where by simple inspection the Jacobian becomes

$$\boldsymbol{G} = \boldsymbol{I} - \left[\frac{1}{2}\widehat{\delta\boldsymbol{\theta}}\right]_\times \,. \tag{2.71}$$

Notice how the reset operation from (2.66) and (2.70) involves quadratic terms (i.e., infinitesimal values), thus its impact on the overall result is minimal and its computation can be neglected similarly to some developments described in previous Sections (e.g., when truncating the Taylor series expansion (2.32) with $n = 1$), however it is here described for the sake of completeness.

2.7 Extended Kalman Filter (EKF)

2.7.1 EKF: Prediction

In this case, the function $f(\cdot)$ and its Jacobians $\boldsymbol{F}_N$ and $\boldsymbol{F}_i$ are drawn from (2.37). The forms of $f(\cdot)$ and $\boldsymbol{F}_N$ depend on the truncation grade we choose—see Sect. 2.5.2 for details. The prediction step is standard EKF,

$$\widehat{\boldsymbol{x}}_t \leftarrow f(\widehat{\boldsymbol{x}}_t, \boldsymbol{u}, \boldsymbol{0}), \tag{2.72a}$$

$$\boldsymbol{P} \leftarrow \boldsymbol{F}_N \boldsymbol{P} \boldsymbol{F}_N^\top + \boldsymbol{F}_i \boldsymbol{Q}_i \boldsymbol{F}_i^\top . \tag{2.72b}$$

The covariance matrix of the perturbation impulses $\boldsymbol{Q}_i$ is defined as in (2.51). The Jacobian of $f(\cdot)$ with respect to the perturbation vector $\boldsymbol{F}_i$ maps these covariances to the state covariance. It can be obtained by first discretizing the system kinematics from (2.28) without removing the noise terms and not considering Δt (Δt is already considered in $\boldsymbol{Q}_i$ and $\boldsymbol{F}_i$ only maps the noise to the corresponding state covariances). The resulting system must be integrated using a truncated Taylor serie expansion in order to compute the Jacobian. For example, with a 1rst-order integration, this perturbation Jacobian is

$$\boldsymbol{F}_i = \begin{bmatrix} \boldsymbol{0} & \boldsymbol{0} & \boldsymbol{0} & \boldsymbol{0} \\ \boldsymbol{I} & \boldsymbol{0} & \boldsymbol{0} & \boldsymbol{0} \\ \boldsymbol{0} & \frac{\partial f_q}{\partial \boldsymbol{\omega}_n} & \boldsymbol{0} & \boldsymbol{0} \\ \boldsymbol{0} & \boldsymbol{0} & \boldsymbol{I} & \boldsymbol{0} \\ \boldsymbol{0} & \boldsymbol{0} & \boldsymbol{0} & \boldsymbol{I} \end{bmatrix} . \tag{2.73}$$

The term $\partial f_q / \partial \boldsymbol{\omega}_n$ is derived from

$$f_q = \boldsymbol{q} \otimes \boldsymbol{q}(\boldsymbol{\omega}_s - \boldsymbol{\omega}_b - \boldsymbol{\omega}_n), \tag{2.74a}$$

$$\approx \boldsymbol{Q}^+(\boldsymbol{q}) \begin{bmatrix} 1 \\ \frac{1}{2}(\boldsymbol{\omega}_s - \boldsymbol{\omega}_b - \boldsymbol{\omega}_n) \end{bmatrix}, \tag{2.74b}$$

which combined with (2.5), we get

$$\frac{\partial f_q}{\partial \boldsymbol{\omega}_n} = -\frac{1}{2} \begin{bmatrix} -q_x & -q_y & -q_z \\ q_w & -q_z & q_y \\ q_z & q_w & -q_x \\ -q_y & q_x & q_w \end{bmatrix} . \tag{2.75}$$

In our EKF, the system covariance matrix $\boldsymbol{P}$ must be initialized with certain values in the diagonal, corresponding to the initial filter standard deviations $\boldsymbol{\sigma}$. However the quaternion block $\boldsymbol{P}(\boldsymbol{q})$ requires a mapping from 3×3 to 4×3 matrix (the orientation standard deviation is specified with 3 angular values). This mapping is done as follows

$$\boldsymbol{P}(\boldsymbol{q}) \leftarrow \boldsymbol{Q}_\theta \boldsymbol{P}(\boldsymbol{q}) \boldsymbol{Q}_\theta^\top , \tag{2.76}$$

where $\boldsymbol{Q}_\theta$ is the Jacobian of the initial state quaternion $\boldsymbol{q}_0$ with respect to initial standard deviations (corresponding to the three angles),

$$\boldsymbol{q}_0 \leftarrow \boldsymbol{q}_0 \otimes \boldsymbol{q}(\sigma_{q_0}^2) = \boldsymbol{Q}^+(\boldsymbol{q}_0)\, \boldsymbol{q}(\sigma_{q_0}^2), \tag{2.77a}$$

$$\approx \boldsymbol{Q}^+(\boldsymbol{q}_0) \begin{bmatrix} 1 \\ \frac{1}{2} \sigma_{q_0}^2 \end{bmatrix}, \tag{2.77b}$$

where combining with (2.5) becomes

$$\boldsymbol{Q}_\theta = \frac{1}{2}\begin{bmatrix} -q_x & -q_y & -q_z \\ q_w & -q_z & q_y \\ q_z & q_w & -q_x \\ -q_y & q_x & q_w \end{bmatrix}. \tag{2.78}$$

2.7.2 EKF: Innovation and Correction

The innovation is obtained as in the ESKF (2.53), with sonar range and 2D velocities, flow or IR range observation Jacobians $\boldsymbol{H}_i = \partial h_i / \partial \boldsymbol{x}_t$, $i \in \{v, \varphi, r\}$. This time differentiating with stardard procedures (2.21) and (2.26) with respect to the true-state $\boldsymbol{x}_t$ instead of the error-state. Appendix 2.B.1 includes some useful derivatives to obtain these Jacobians. We perform the same χ^2-test outlier rejection explained for the ESKF and the filter correction follows the standard EKF formulation,

$$\boldsymbol{K} = \boldsymbol{P}\boldsymbol{H}_i^\top \boldsymbol{Z}^{-1}, \tag{2.79a}$$

$$\widehat{\boldsymbol{x}}_t \leftarrow \widehat{\boldsymbol{x}}_t + \boldsymbol{K}\boldsymbol{z}, \tag{2.79b}$$

$$\boldsymbol{P} \leftarrow \boldsymbol{P} - \boldsymbol{K}\boldsymbol{Z}\boldsymbol{K}^\top. \tag{2.79c}$$

Notice that, unlike the ESKF updates (2.57), the sum in (2.79b) implies that the orientation escapes the $\mathcal{SO}(3)$ manifold, and thus that quaternion re-normalization is required, as explained in the following.

2.7.3 EKF: Reset Operation

When using the EKF, the state quaternion needs to be normalized after the correction step. Then, the respective block of the covariance matrix $\boldsymbol{P}(\boldsymbol{q})$, should be mapped accordingly by the Jacobian of the normalization operation $\boldsymbol{G}_n$, with

$$\boldsymbol{P}(\boldsymbol{q}) \leftarrow \boldsymbol{G}_n \boldsymbol{P}(\boldsymbol{q}) \boldsymbol{G}_n^\top. \tag{2.80a}$$

The Jacobian of this normalization operation $\boldsymbol{G}_n$ is detailed in the following.

2.7.3.1 EKF: Reset Jacobian

After a filter correction in EKF the quaternion needs to be normalized, and so the system covariance matrix reset accordingly to the Jacobian of this operation. The quaternion normalization operation consists on

$$\boldsymbol{q}_n = \frac{\boldsymbol{q}}{||\boldsymbol{q}||}, \tag{2.81}$$

Then its partial derivative with respect to the quaternion elements is

$$\boldsymbol{G}_n = \begin{bmatrix} \frac{q_x^2+q_y^2+q_z^2}{||\boldsymbol{q}||^3} & -\frac{q_w q_x}{||\boldsymbol{q}||^3} & -\frac{q_w q_y}{||\boldsymbol{q}||^3} & -\frac{q_w q_z}{||\boldsymbol{q}||^3} \\ -\frac{q_w q_x}{||\boldsymbol{q}||^3} & \frac{q_w^2+q_y^2+q_z^2}{||\boldsymbol{q}||^3} & -\frac{q_x q_y}{||\boldsymbol{q}||^3} & -\frac{q_x q_z}{||\boldsymbol{q}||^3} \\ -\frac{q_w q_y}{||\boldsymbol{q}||^3} & -\frac{q_x q_y}{||\boldsymbol{q}||^3} & \frac{q_w^2+q_x^2+q_z^2}{||\boldsymbol{q}||^3} & -\frac{q_y q_z}{||\boldsymbol{q}||^3} \\ -\frac{q_w q_z}{||\boldsymbol{q}||^3} & -\frac{q_x q_z}{||\boldsymbol{q}||^3} & -\frac{q_y q_z}{||\boldsymbol{q}||^3} & \frac{q_w^2+q_x^2+q_y^2}{||\boldsymbol{q}||^3} \end{bmatrix}, \tag{2.82}$$

which for the sake of clarity and simplicity its development has been avoided.

2.8 Observability Analysis

The observability analysis of the system needs the evaluation of the rank and continuous symmetries of the observability matrix defined from the Lie derivatives [24]. Following this work, we detect three continuous symmetries, corresponding to the non-observable modes of *xy* translation, and rotation around the direction of gravity (i.e., the *yaw* angle),

$$\begin{aligned} \boldsymbol{\omega}_s^1 &= [1, 0, 0, 0, \cdots, 0], \\ \boldsymbol{\omega}_s^2 &= [0, 1, 0, 0, \cdots, 0], \\ \boldsymbol{\omega}_s^3 &= [-p_y, p_x, 0, -v_y, v_x, 0, -\frac{q_z}{2}, -\frac{q_y}{2}, \frac{q_x}{2}, \frac{q_w}{2}, 0, \cdots], \end{aligned}$$

where $\{p_x, p_y, v_x, v_y, q_x, q_y, q_z, q_w\}$ are position, velocity and quaternion components. All other modes, including all biases, are observable as long as the maneuvers performed span the observable directions. The limitations on maneuverability imposed by the MAV dynamics have a negative impact on the observability of certain modes, in particular on the accelerometer bias in the *xy* axes, and the gyrometer bias in the *z* axis. These biases are observable only when the MAV escapes the hovering attitude, and their convergence increases the further we deviate from hovering, as shown in the experiments section—see Sect. 2.10. Therefore, it is beneficial to drive the MAV in aggressive maneuvers. This requires a flight controller with good stability conditions away from the hovering situation, such as the one we present hereafter.

2.9 Control and Planning

We validate the previous state estimation methods with a quadrotor, given its mechanical simplicity [27] and ease of control. This section describes its dynamical model and the chosen control scheme.

2.9.1 Dynamic Model

Quadrotors are typically equipped with four aligned coplanar propellers. Motion control is achieved by altering the rotation speed of these propellers, thereby changing its torque load and thrust lift characteristics—see Fig. 2.4.

Let us consider a global coordinate frame w, assumed inertial and defined by unitary column vectors $[{}^{w}\boldsymbol{x}, {}^{w}\boldsymbol{y}, {}^{w}\boldsymbol{z}]$, and a body reference frame b, defined also by $[{}^{b}\boldsymbol{x}, {}^{b}\boldsymbol{y}, {}^{b}\boldsymbol{z}]$ and centered in the center of mass of the vehicle. The dynamic model of the vehicle can be expressed as

$$\dot{\boldsymbol{p}} = \boldsymbol{v}\,, \tag{2.83a}$$

$$m\boldsymbol{a} = -f\,\boldsymbol{R}\,{}^{w}\boldsymbol{z} + m\,\boldsymbol{g}\,, \tag{2.83b}$$

$$\dot{\boldsymbol{R}} = \boldsymbol{R}[\boldsymbol{\omega}]_{\times}\,, \tag{2.83c}$$

$$\boldsymbol{\tau} = \mathcal{I}\dot{\boldsymbol{\omega}} + \boldsymbol{\omega} \times \mathcal{I}\boldsymbol{\omega}, \tag{2.83d}$$

where $m \in \mathbb{R}$ is the mass, $\mathcal{I} \in \mathbb{R}^{3\times3}$ is the inertia matrix with respect to the body frame, and ${}^{w}\boldsymbol{z} \triangleq \begin{bmatrix}0 & 0 & 1\end{bmatrix}^{\top}$. The control inputs of the plant are the total thrust $f \in \mathbb{R}$, and the total moment $\boldsymbol{\tau} = \begin{bmatrix}\tau_1 & \tau_2 & \tau_3\end{bmatrix}^{\top} \in \mathbb{R}^3$ along all axes of the body-fixed frame. The dynamics of rotors and propellers are neglected and it is assumed that the force f_i of each propeller is directly controlled. The total thrust, $f = \sum_{j=1}^{4} f_j$, acts in the direction of the z axis of the body-fixed frame, which is orthogonal to the plane defined by the centers of the four propellers. The relationship between the single motor forces f_i, the total thrust f, and the total moment $\boldsymbol{\tau}$, can be written as

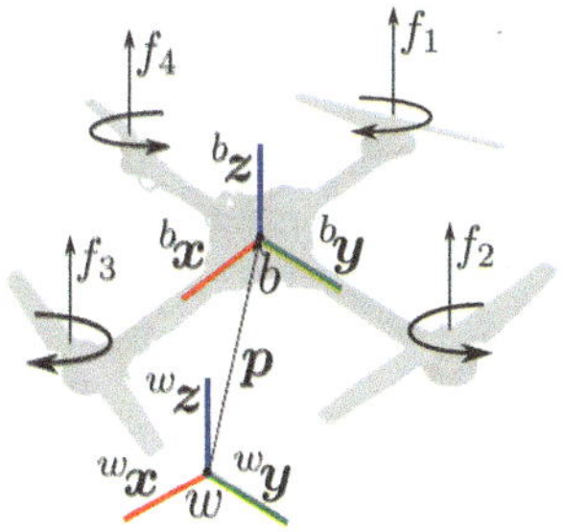

Fig. 2.4 Quadrotor scheme with reference frames, thrust vectors and propeller rotation directions

$$\begin{bmatrix} f \\ \tau_1 \\ \tau_2 \\ \tau_3 \end{bmatrix} = \begin{bmatrix} 1 & 1 & 1 & 1 \\ 0 & -d & 0 & d \\ d & 0 & -d & 0 \\ -c & c & -c & c \end{bmatrix} \begin{bmatrix} f_1 \\ f_2 \\ f_3 \\ f_4 \end{bmatrix}, \tag{2.84}$$

where c is a constant value and d is the distance from the center of mass (b) to a rotor axis, considering all rotors equidistant. For non-zero values of d, (2.84) can be inverted, therefore our assumption that f and $\boldsymbol{\tau}$ are the inputs of the plant is valid.

2.9.2 Position and Attitude Controllers

We want to control the quadrotor with desired positions, heading, linear velocities and accelerations (i.e., $\boldsymbol{p}_d$, ψ_d, $\boldsymbol{v}_d$ and $\boldsymbol{a}_d$) with a controller design based on the nonlinear tracking controller developed on the special Euclidean group $\mathscr{SE}(3)$ [15]. For this, the quadrotor control inputs f, $\boldsymbol{\tau}$ from (2.84) (see Fig. 2.1) are chosen as

$$f = -\left(-k_p \boldsymbol{p}_e - k_v \boldsymbol{v}_e - m\,\boldsymbol{g} + m\boldsymbol{a}_d\right) \boldsymbol{R}^{w}\boldsymbol{z}\,, \tag{2.85a}$$

$$\boldsymbol{\tau} = -\,k_\theta \boldsymbol{\theta}_e - k_\omega \boldsymbol{\omega}_e + \boldsymbol{\omega} \times \mathscr{I}\boldsymbol{\omega} - \mathscr{I}\left([\boldsymbol{\omega}]_\times \boldsymbol{R}^\top \boldsymbol{R}_c \boldsymbol{\omega}_c - \boldsymbol{R}^\top \boldsymbol{R}_c \dot{\boldsymbol{\omega}}_c\right), \tag{2.85b}$$

with $k_p, k_v, k_\theta, k_\omega$ positive definite gains to be tuned. $\boldsymbol{p}_e$, $\boldsymbol{v}_e$, $\boldsymbol{\theta}_e$ and $\boldsymbol{\omega}_e$ are the position, velocity, orientation and angular rate errors, defined by

$$\boldsymbol{p}_e = \boldsymbol{p} - \boldsymbol{p}_d\,, \tag{2.86a}$$

$$\boldsymbol{v}_e = \boldsymbol{v} - \boldsymbol{v}_d\,, \tag{2.86b}$$

$$\boldsymbol{\theta}_e = \frac{1}{2}\left[\boldsymbol{R}_c^\top \boldsymbol{R} - \boldsymbol{R}^\top \boldsymbol{R}_c\right]^\times, \tag{2.86c}$$

$$\boldsymbol{\omega}_e = \boldsymbol{\omega} - \boldsymbol{R}^\top \boldsymbol{R}_c \boldsymbol{\omega}_c\,. \tag{2.86d}$$

$\boldsymbol{R}_c$ and $\boldsymbol{\omega}_c$ are the internally controlled orientation and angular velocity, as produced by the position controller, refer to [15] for more details on their definitions. The symbol $[\cdot]^\times$ represents the map $\mathfrak{so}(3) \to \mathbb{R}^3$, which is the inverse operation of $[\cdot]_\times$.

Using this controller, if the initial attitude error is less than 90°, the zero equilibrium of the tracking errors is exponentially stable, i.e. $\left[\boldsymbol{p}_e\ \boldsymbol{v}_e\ \boldsymbol{\theta}_e\ \boldsymbol{\omega}_e\right] \to \boldsymbol{0}$. Furthermore, if the initial attitude error is between 90° and 180°, then the zero equilibrium of the tracking errors is almost globally exponentially attractive. The reader can refer to [15] for convergence and stability analysis and to [26] for experimental results.

2.9.3 Trajectory Planning

With the planning module (see Fig. 2.1) we generate trajectories in Cartesian space. These trajectories consist of the desired values fed to the controller above, $\boldsymbol{p}_d$, ψ_d, $\boldsymbol{v}_d$ and $\boldsymbol{a}_d$. Our planner design is based on [26] which guarantees dynamically feasible trajectories by proving that our dynamic system (2.83) is differential flat [6]. This means that our dynamic system can be formulated as an algebraic function of the flat outputs, which are

$$\eta = \left[\boldsymbol{p}^\top \; \psi\right] , \tag{2.87}$$

or their derivatives. These algebraic relations involve the fourth derivative of the position $\boldsymbol{p}$, called snap, and the second derivative of the heading ψ [26]. Therefore, to generate smooth and feasible 3D trajectories, it is convenient and sufficient to minimize this snap using the following cost functional

$$\min \int_{t_0}^{t_f} \left(\mu_p \left\| \frac{\partial^4 \boldsymbol{p}(t)}{\partial t^4} \right\|^2 + \mu_\psi \left\| \frac{\partial^2 \psi(t)}{\partial t^2} \right\|^2 \right) dt , \tag{2.88}$$

where μ_p and μ_ψ are tuning parameters, subject to the desired boundary conditions on the flat outputs and their concerned derivatives. This minimization problem can be formulated as a quadratic program [26], also including intermediate waypoints.

2.10 Validation and Experimental Results

In order to study the performances and limitations of the proposed state estimation setup, we first present experiments with synthetic data under realistic flight conditions.

2.10.1 Simulation Results

We generate ground truth trajectories that account for quadrotor motion constraints (i.e., under-actuated vehicle with only 4 DoF). We subsequently generate corrupted sensor measurements, with noise and bias characteristics similar to those of the real units, delivering data synchronously at 100 Hz, and taking advantage of a MATLAB toolbox (checkout the on-line software[1] for more details on quadrotor dynamic values and sensor parameters). For the benefit of the community, we also make the MATLAB

[1] https://gitlab.iri.upc.edu/asantamaria/QuadSim.

odometry estimation code[2] public. The optimized high-rate C++ implementation is available upon request.

We validate our method comparing the produced estimates with respect to precise ground truth measurements. Notice that we do not compare our performances against the more sophisticated VIO algorithms for the reasons exposed in the introduction of this chapter, namely the lack of key-frames and lengthy feature tracks in our estimation pipeline.

In a first set of validations we simulate smart camera sonar readings and 2D linear velocities—see Sect. 2.4.2. Figure 2.5 shows the results using an ESKF with GE, Q0B and $\boldsymbol{F}_1$. The 60s trajectory encompasses an initial phase of hovering (phase A); a forward movement along the x axis (B); a backward movement (C) to the initial position; a second hovering phase (A); and finally left (D) and then right (E) movements along the y axis. The true trajectory finishes precisely at the starting position. x-y-z magnitudes are colored R-G-B, and $\pm 3\sigma$ error bounds are also plotted.

The xy position is not observable, and its estimated error increases with time (top frame in Fig. 2.5). However, the drift is small, a few centimeters after the whole one-minute flight. Altitude and velocity estimates converge quickly thanks to the smart camera measurements. Notice the transient increase of the position and velocity errors (phases B and D). This is due to the uncertainty in *yaw* (blue track of third plot in Fig. 2.5, not observable) which produces position errors perpendicular to the displacement vector. These errors decrease as the MAV returns to the origin (C and E).

One of the most interesting aspects is the marginal observability of some of the biases. On one hand, an acceleration bias becomes observable when the quadrotor orientation produces a non-null projection of the gravity vector over its axis. This makes the accelerometer's z-bias readily observable, but x- and y-biases require tilting the rotor plane. This is observable in Fig. 2.5, plot 'Acc. Bias'. This is a critical issue due to the quadrotor's restricted dynamics, seeking always horizontality and thus making these observations difficult. To further investigate this effect, the impact of a rotation over y (with respect to x) on the observability of $\boldsymbol{a}_b(x)$ (with respect to $\boldsymbol{a}_b(y)$) is illustrated in Fig. 2.6. A quadrotor can easily tilt to 20°, but holding this tilt for a period of seconds represents a sustained high acceleration that is usually not desirable. Otherwise, this bias should be estimated beforehand and set up in the filter.

On the other hand, The gyroscope's *pitch* and *roll* biases are well observable thanks to the gravity vector. The *yaw* bias $\boldsymbol{\omega}_b(z)$ (blue in 'Ang. Vel.' plot) becomes observable only after a very long period, by taking small profit of transient inclinations that bring the gravity vector away from the local z axis. To achieve a better performance in *yaw*-bias, it is advisable to use extra heading measurements such as a compass or other stable external references.

[2]https://gitlab.iri.upc.edu/asantamaria/QuadOdom.

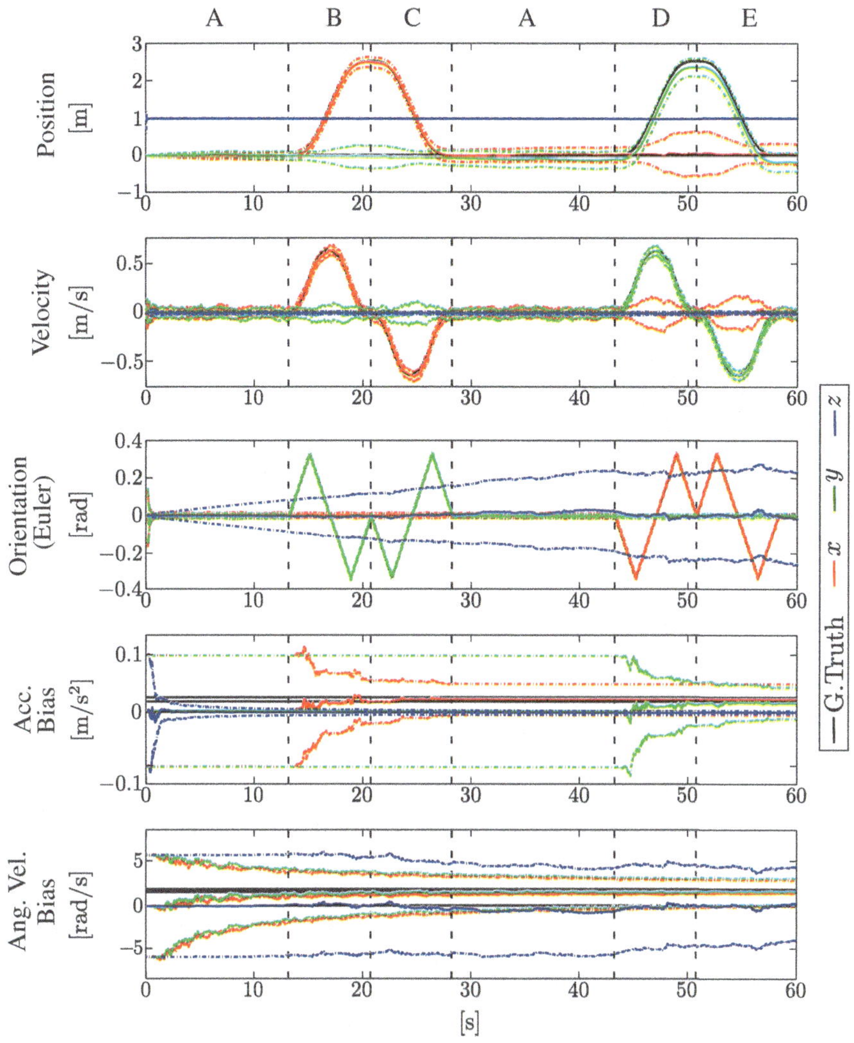

Fig. 2.5 Simulation trajectories using IMU, 2D linear velocities and sonar range]Trajectories of all the estimated states, with their $\pm 3\,\sigma$ bounds, for a simulation experiment using IMU, and 2D linear velocities and sonar readings from the smart camera. The vertical sections indicate the hovering (A), forward (B), backward (C), left (D) and right (E) maneuvers

In a second set of validations we benchmark all filter types using the same scenario but with the simulated quadrotor equipped with an IMU, a smart camera producing flow data and a range sensor with measurements as those of an IR time-of-flight ranger,—see Sects. 2.4.3 and 2.4.4.

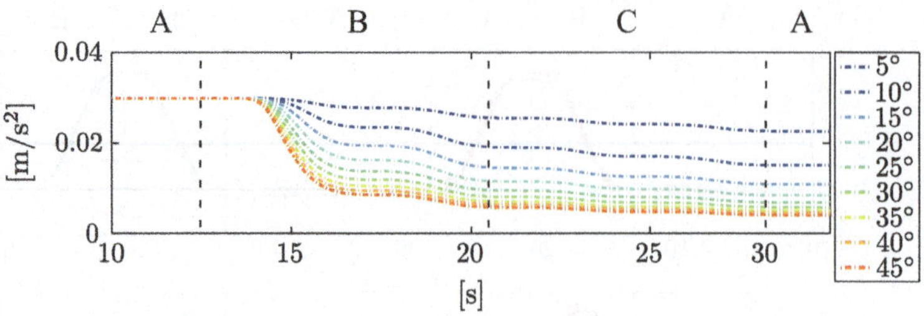

Fig. 2.6 Evolution of the estimated error of the accelerometer bias in the x (y) axis for different *pitch* (*roll*) angles

2.10.1.1 Position RMSE and Orientation Error Evaluation

To analyze the resulting filter estimations we perform N trajectory simulations. We evaluate the Root Mean Square Error (RMSE) between each component i of the estimated vehicle positions (x,y,z) with respect to ground truth, for all time steps k

$$\epsilon_i = \sqrt{\frac{1}{N\,s}\sum_{j=1}^{N}\sum_{k=1}^{s}(p_{i,k} - \hat{p}^j_{i,k})^2}\,, \tag{2.89}$$

where s is the number of time samples of each experiment, $p_{i,k}$ is the i-th component of the true vehicle position at time k, and $\hat{p}^j_{i,k}$ is its estimate, computed by the filter, corresponding to the j-th among N simulated trajectories.

To analyze the orientation error we use as in [20], which in turn is based on [4], the orientation error metric defined as

$$\varsigma = \frac{1}{2}tr(\boldsymbol{I} - \boldsymbol{R}^\top\widehat{\boldsymbol{R}}) \quad \in \mathbb{R}\,, \tag{2.90}$$

where $\boldsymbol{R}$ and $\widehat{\boldsymbol{R}}$ are respectively the ground truth and estimated vehicle orientations.

Table 2.2 shows both position RMSE and the abovementioned orientation error metric (no units) achieved at the end of $N = 20$ simulated flights of almost 10 min and 500 m each, performing representative movements (e.g., up/down, forward/backward, left/right). For the sake of simplicity only some of the filter variants are reported, and to ease the comparison some filter characteristics are colored. The results in Table 2.2 show that there is no significant performance difference between filter designs.

2.10.1.2 Average NEES Evaluation

A recursive estimator is consistent when the estimation errors are zero-mean and have covariance matrix equal to that reported by the estimator. To evaluate the consistency

Table 2.2 Estimation error statistics after 10 min flights of 500 m in straight line. Root Mean Squared Error (RMSE) over 20 experiments for Cartesian position elements (x, y, z) and rotation error index ς at the end of the trajectory. Color in the filter variant names are added for comparison purposes (those variants with the same color only differ from the colored characteristic)

Error component ϵ_i	Filter variant										
	EKF	EKF	EKF	EKF	EKF	ESKF	ESKF	ESKF	ESKF	ESKF	ESKF
	$\boldsymbol{F}_1$	$\boldsymbol{F}_1$	$\boldsymbol{F}_1$	$\boldsymbol{F}_2$	$\boldsymbol{F}_3$	$\boldsymbol{F}_1$	$\boldsymbol{F}_1$	$\boldsymbol{F}_1$	$\boldsymbol{F}_2$	$\boldsymbol{F}_3$	$\boldsymbol{F}_1$
	Q0F	Q0B	Q1	Q1	Q1	Q0F	Q0B	Q1	Q1	Q1	Q0F
	LE	LE	LE	LE	LE	GE	GE	GE	GE	GE	LE
x [m]	10.54	10.48	10.30	10.26	10.26	10.58	10.37	10.13	10.12	10.12	10.38
y [m]	11.13	11.07	10.85	10.81	10.81	11.00	10.82	10.55	10.58	10.58	10.91
z [mm]	7	6	7	6	6	7	7	7	7	7	7
ς $(\cdot 10^{-3})$	2	2	2	2	2	2	2	2	2	2	2

of the filters we use, as in [43] which in turn is based in [1], the Average Normalized Estimation Error Squared (ANEES) for N Monte Carlo runs, defined as

$$\eta_k = \frac{1}{N} \sum_{j=1}^{N} \left(\boldsymbol{B}_k - \widehat{\boldsymbol{B}}_k^j \right)^\top \boldsymbol{P}_k^{j\,-1} \left(\boldsymbol{B}_k - \widehat{\boldsymbol{B}}_k^j \right) , \tag{2.91}$$

where $\boldsymbol{B}_k$ is the 6 DoF true body pose at time k (i.e., ground truth) and $\mathcal{N}(\overline{\boldsymbol{B}}_k^j, \boldsymbol{P}_k^j)$ is its Gaussian estimate, obtained by filtering, corresponding to the j-th among the N Monte Carlo runs. Each run is done with a different seed for the random generator affecting the process noises and the measurement noises. We now can compute the double-sided 95% probability concentration region, which, for 6 DoF and $N = 25$ runs, has the upper and lower bounds given by

$$\overline{\eta} = \frac{\chi^2_{(25\times 6)}(1 - 0.975)}{25} = 7.432 , \tag{2.92a}$$

$$\underline{\eta} = \frac{\chi^2_{(25\times 6)}(1 - 0.025)}{25} = 4.719 . \tag{2.92b}$$

If $\eta_k < \underline{\eta}$ for a significant amount of time (more than 2.5% of the time), the filter is considered conservative. Similarly, if $\eta_k > \overline{\eta}$ (also by more than 2.5% of the time), the filter is considered optimistic and therefore inconsistent. Figure 2.7 shows an example of the ANEES for the 6 DoF body frame pose $\left[x \; y \; z \; \phi \; \theta \; \psi\right]^\top$ over 25 runs of the same experiment ($N = 25$). We estimated the pose using the two extreme filter variants in terms of simplicity, an EKF with $\boldsymbol{F}_1$ and Q0B options; and an ESKF with GE, $\boldsymbol{F}_3$ and Q1 options. The gray horizontal band between abscissas mark the 95% consistency region with $\overline{\eta} = 4.719$ and $\underline{\eta} = 7.432$. Both filter variants are shown to

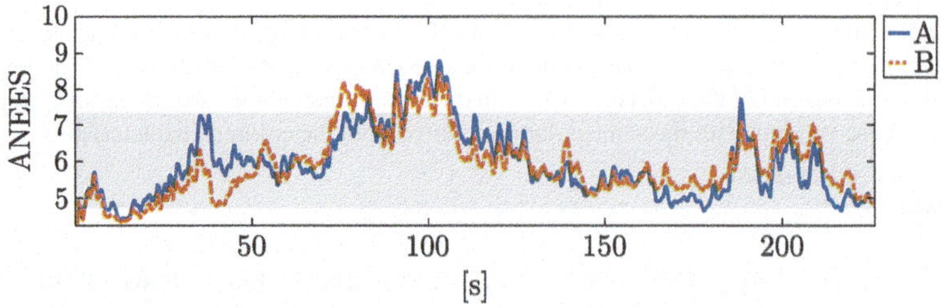

Fig. 2.7 Average Normalized Estimation Error (ANEES) of the 6 DoF body frame pose $[x\ y\ z\ \phi\ \theta\ \psi]^\top$ over 25 runs of the same experiment. The filter variants are: (A) EKF with $\boldsymbol{F}_1$ and Q0B; and (B) ESKF with GE, $\boldsymbol{F}_3$ and Q1 options. Note how the ANEES is normalized and does not need to differentiate between position and orientation. The gray horizontal band between abscissas $\overline{\eta} = 4.719$ and $\underline{\eta} = 7.432$ mark the 95% consistency region. Both filter propagation and updates are running at 100Hz

be neither conservative nor inconsistent—see the on-line software simulator for all involved parameters during simulation and estimation.

2.10.2 Experimental Results

We divided the experiments in two sets depending on the hardware setting used:

- **Setting A**: ASCTEC Pelican[3] research platform shown in Fig. 2.8, equipped with a MicroStrain 3dm-gx3-25 IMU running at 100 Hz and a PX4-OpticalFlow smart camera [10] with a rate of 200 Hz. We associate each PX4 measurement with the closest IMU measurement. The sensors are attached below the platform using silicon damping links to reduce motor vibrations. In these experiments the IMU accelerometers and gyroscopes, the sonar range and the 2D linear velocities are used in the filtering process. Here, the vehicle is driven manually only to validate the method by comparing the estimated state with the one obtained from an Optitrack[4] motion capture system. Moreover, we show results for state estimation driving the vehicle in a GPS denied zone. Experiments using setting A are presented in Video 1, referenced in Appendix 5.A.
- **Setting B**: ASCTEC Hummingbird[5] research platform shown in Fig. 2.8. This platform has an off-the-shell built-in IMU (i.e., 3-axis accelerometer and 3-axis gyroscope) running at 100 Hz, and we equipped it with a PX4-OpticalFlow smart camera with a rate of 200 Hz and a TeraRangeOne IR range sensor [35] with a frequency up to 800 Hz. All algorithms for both odometry estimation and control are running on board within an Odroid XU3 board running Ubuntu 14.04LTS and

[3] www.asctec.de/en/uav-uas-drones-rpas-roav/asctec-pelican.

[4] www.optitrack.com.

[5] www.asctec.de/en/uav-uas-drones-rpas-roav/asctec-hummingbird.

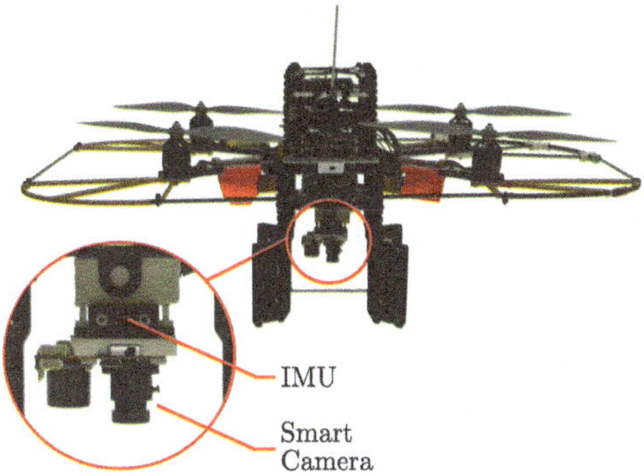

(a) ASCTEC Pelican quadrotor equipped with a Microstrain 3dm-gx3-25 IMU and a PX4 optical flow smart camera (setting A).

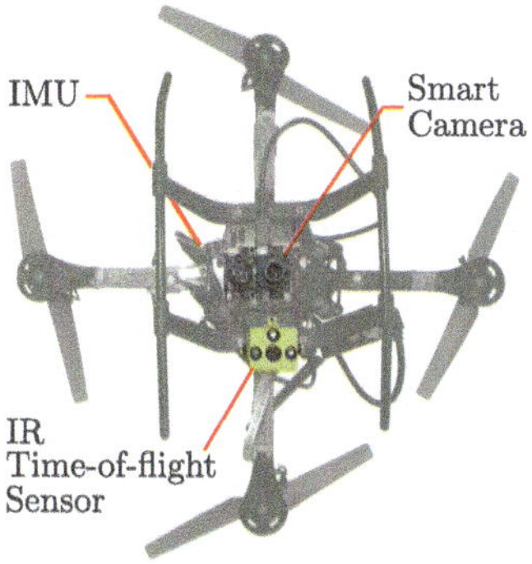

(b) Bottom view of an ASCTEC Hummingbird quadrotor equipped with a built-in IMU, a PX4 optical flow smart camera and an IR time-of-flight range sensor (setting B).

Fig. 2.8 Quadrotors used in the experiments to test the odometry estimation approach

ROS Indigo. Although the Odroid XU3 board has a Cortex^{TM}-A7 quad core CPU, all algorithms are using less than half the cores. Finally, a Qualisys[6] motion capture system running at 100 Hz has been used for ground-truth comparison. Experiments using setting B are presented in Video 1, referenced in Appendix 5.A.

2.10.2.1 Experiments with Setting A

This set of experiments consists on manually flying several times the platform setting A at the Institut de Robòtica i Informàtica Industrial (IRI) testbed. Table 2.3 shows all non-null sigma values for the filter tuning using hardware setting A. All filter variants use the same parameters. For the IMU noises, we followed the datasheet to obtain $\sigma_{a_n} = 5.3\,mm/s^2$ and $\sigma_{\omega_n} = 3.6\,rad/s$. However, empirical IMU data has revealed that the mechanical damping is not sufficient, and propeller vibrations inject much higher noises to the acceleration measurements. For this reason, the value of σ_{a_n} is significantly increased as shown in the table. We assume null bias random walks, as our flights are short enough. Initial uncertainties are mostly null, except for altitude $\sigma_{\delta z}$, initial roll $\sigma_{\delta\phi}$ and pitch $\sigma_{\delta\theta}$, accelerometer bias $\sigma_{\delta a_b}$, and gyrometer roll and pitch bias $\sigma_{\delta\omega_b}$.

The PX4 sensor has two key limitations that need to be addressed algorithmically in order to improve its robustness and usability. The first one is its inability to measure altitudes below 30 cm. For this, we divide each experiment in 4 phases depending on the robot state: landed (A), taking-off (B), flying (C) or landing (D). Thus during A,

[6] www.qualisys.com.

Table 2.3 Filter tuning parameters (std; σ symbols omitted) using setting A (IMU accelerometers and gyroscopes, and PX4 sonar range and 2D linear velocities)

Sensor noises				Initial std. dev.			
$\boldsymbol{a}_n$ [m/s^2]	$\boldsymbol{\omega}_n$ [rad/s]	z_{of} [m]	v_{of} [m/s]	δz [m]	$\delta\phi, \delta\theta$ [rad]	$\delta \boldsymbol{a}_b$ [m/s^2]	$\delta\boldsymbol{\omega}_b$ [rad/s]
0.4	0.005	0.05	0.1	0.05	0.05	0.02	0.004

B and D intervals, the PX4 output is not reliable. We address this problem by reading status data of the robot to acknowledge whether it is in flying mode or not (e.g., motors on/off). Before take-off (phase A), we overwrite the optical flow measurements by assuming that the MAV is on zero position, with zero velocity, and we set a small observation covariance. Thus, the bias uncertainties on the z acceleration and x and y angular velocities are reduced. If the robot is flying below the minimum altitude of 0.3 m (phases B and D), detected with PX4 readings, we set high PX4 covariances because the measures are not trustable. Hence the filter is propagating the nominal state with the IMU dynamical model, and practically does not correct with the PX4 sensor. During regular flight (phase C) the observation covariances are set to those in Table 2.3, thus allowing the PX4 to correct the estimations properly.

A last feature of our algorithm deals with the second PX4 data integrity issue. In some occasions, tilts over 20° produce wrong sonar echoes, deriving in aberrant PX4 outputs. We detect altitude outliers using the modified Thompson Tau method [45]. Once an outlier is detected, the filter correction can be accomplished by using the optical flow measurement as a function of the IMU's gyrometer measurement $\boldsymbol{\omega}_s$ and the state variables, substituted in (2.25), with $\boldsymbol{n}_\varphi$ a one-pixel Gaussian noise. For simplicity, we however re-used the observation model (2.14) by computing the velocity $\boldsymbol{v}_{of}$ with (2.16) substituting $h_{of} \leftarrow \boldsymbol{p}(z)$ and $\boldsymbol{\omega}_{of} \leftarrow \boldsymbol{\omega}_s$. In this case, to compensate for the correlation between the state and the measurement, we increased the sensor covariances $\boldsymbol{N}_v$ slightly. Figure 2.9 shows the reconstructed $\boldsymbol{v}_{of}$ in front of typical outliers. With this improvement of PX4 usability we perform the manual flights in an outdoor scenario corresponding to a GPS-denied zone in which ground truth from external positioning systems is not available. Instead, we drive the platform around some fixed obstacles, taking-off and landing on the same base point.

Figure 2.10a shows the calibrated outdoor scenario with the take-off and landing platform in the center, surrounded by vertical cylindric obstacles. Figure 2.10b shows the estimated trajectories of 15 flights of approximately two minutes and 70 m each (using an ESKF with GE, Q1 and $\boldsymbol{F}_3$). The quadrotor is driven manually around the obstacles, which results in different flight paths. In almost all results, the estimated trajectory does not touch any obstacles or walls. The final estimated land points have a standard deviation of $\sigma = [0.50\ 0.53\ 0.01]$ m from the center of the landing base. The orange ellipse in Fig. 2.10b corresponds to the zone of confidence of 95% probability of this landing point distribution. All true landings were done inside the base area which measures 0.4×0.4 m. The orange ellipse is thus the composition of the landing error and the estimation error. The blue line corresponds to a sample

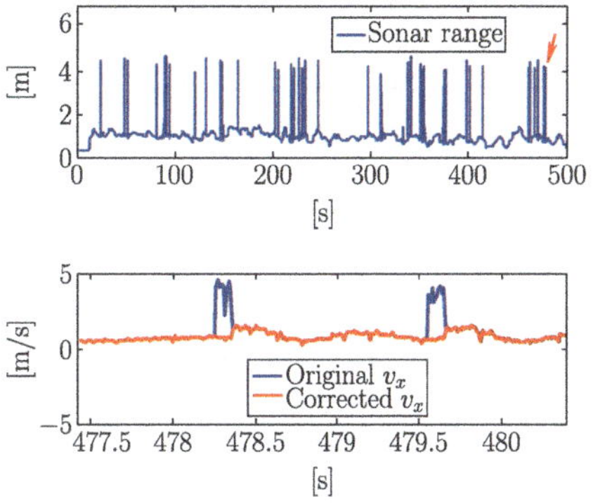

Fig. 2.9 Sonar range outlier detection and correction of PX4 2D velocity measurements. Top-left: full 500 s altitude sequence with numerous outliers. Other plots: zoom of the arrowed outliers in top plot, showing original and corrected altitude and velocity measurements

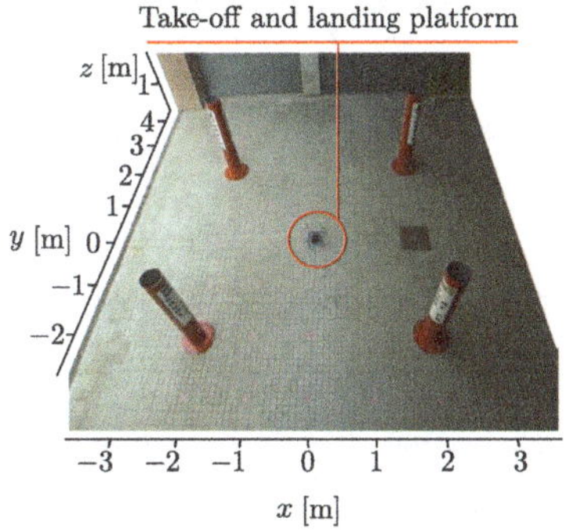

(a) Calibrated outdoor flying arena for real robot testings. The quadrotor takes off and lands at the same point in the middle of the field (base).

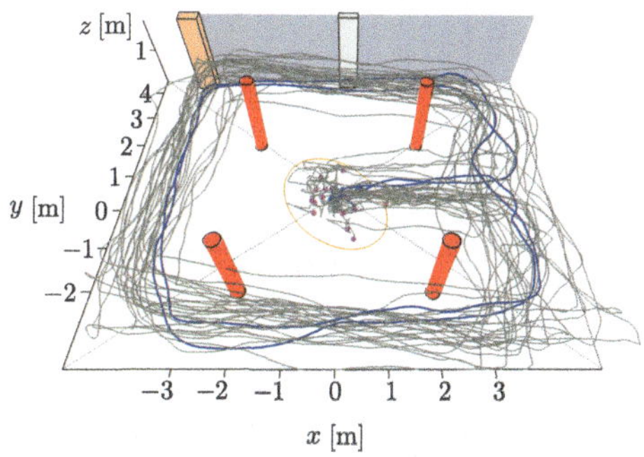

(b) Trajectory estimations for 15 different flights (2 minutes, 70 m each) using setting A. A sample trajectory is shown in blue. The orange ellipse corresponds to the 95% confidence region for the landing point.

Fig. 2.10 Experiments using setting A in an outdoor GPS-denied scenario

flight whose specific results are shown in Fig. 2.11. Notice in the zoomed details of position and velocity plots how the estimation errors increase during take-off and landing periods due to the PX4 limitations explained above. The transition to observability B→C once the PX4 data is recovered is also visible.

2.10.2.2 Experiments with Setting B

This set of experiments consists on executing autonomously several trajectories (i.e., the control part uses only the state estimation as input and including take-off and landing maneuvers) in the PERCH lab (Penn Engineering Research Collaborative Hub) at the University of Pennsylvania indoor testbed.

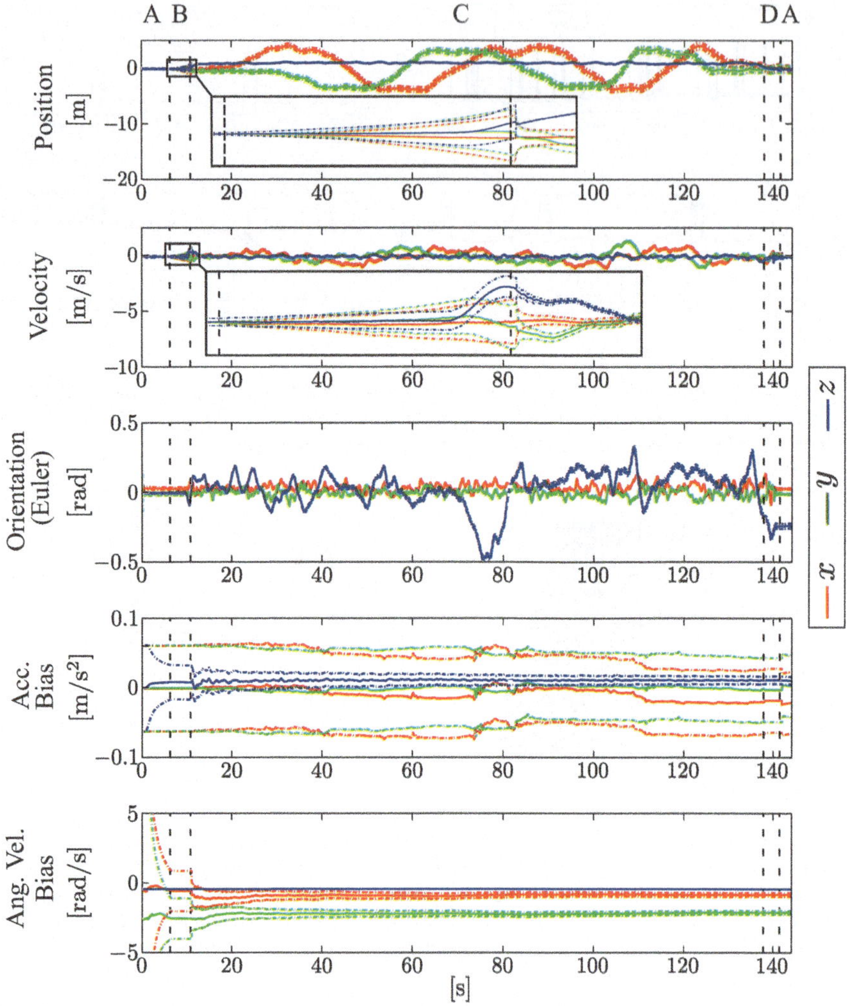

Fig. 2.11 Trajectories of all the estimated states, with their $\pm 3\,\sigma$ bounds, for the outdoors experiment using setting A. The vertical sections indicate the landed states (A), and the take-off (B), flight (C) and landing (D) maneuvers. Zooms are provided to appreciate the transition between phases B→C

Table 2.4 shows all non-null sigma values for the filter tuning using hardware setting B. All filter variants use the same parameters.

Figure 2.12a and b show the on board state estimates compared to measurements from a Qualisys motion capture system, for both positioning and orientation in a sample experiment. As detailed in previous sections, the height of the platform (i.e., z axis in Fig. 2.13c) is observable thanks to the range measurement, thus its error is low. Similarly, roll and pitch estimation errors are low due to the observability of the

Table 2.4 Filter tuning parameters (std; σ symbols omitted) using setting B (IMU accelerometers and gyroscopes, IR ranger and PX4 flow)

Sensor noises				Initial std. dev.			
$\boldsymbol{a}_n$ [m/s^2]	$\boldsymbol{\omega}_n$ [rad/s]	${}^{c}z$ [m]	$\boldsymbol{\phi}$ [pix/s]	δz [m]	$\delta\phi$, $\delta\theta$ [rad]	$\delta\boldsymbol{a}_b$ [m/s^2]	$\delta\boldsymbol{\omega}_b$ [rad/s]
0.4	0.005	0.02	10	0.05	0.05	0.02	0.004

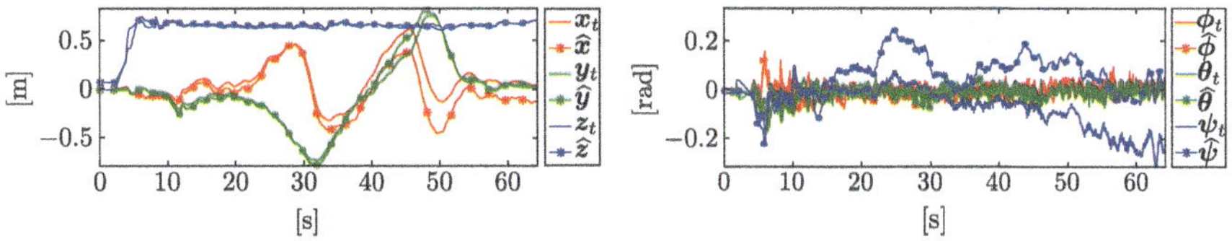

(a) Position (b) Orientation

Fig. 2.12 Comparison between the estimation of a sample trajectory (using an ESKF with GE, $\boldsymbol{F}_3$ and Q1) and ground-truth (Qualisys motion capture system). The corresponding position RMSE is [0.130, 0.051, 0.094] and the error STD is [0.087, 0.050, 0.032]

gravity direction provided by the fused IMU data. Finally, the xy errors grow with time, partly because of the integration of noisy xy velocities, but mostly due to the effect that an unobserved yaw angle ψ has on translation errors.

Figure 2.13 shows experiments for two different trajectories, Fig. 2.13a and b. We launched 25 autonomous runs for each trajectory with a desired height of 1 m and maximum cruise velocity around 1 m/s (notice the superposition of the estimated and ground-truth trajectories in blue and gray respectively). The error statistics for all runs in terms of RMSE are shown in Fig. 2.13c. Using similar trajectories we also pushed the smart camera to its limits, by increasing the maximum cruise velocity, and we reached 2.5 m/s flying at 1.5 m height without significant increase in the resulting estimation and control performance.

In order to show the viability of the proposed methods to drive autonomously the vehicle during realistic flight durations, we performed long experiments consisting on continuous trajectory loops during almost 10 min (i.e., a full battery discharge). Figure 2.14 shows a comparison between the estimated ($\hat{\boldsymbol{p}}$) and ground-truth ($\boldsymbol{p}_t$, obtained with a Qualisys motion capture system) trajectories for one of these experiments with a position RMSE of [0.47, 0.67, 0.035] m, and standard deviation [0.29, 0.48, 0.003] m. The maximum position error at the end of the flight is [0.73, 1.65, 0.028] m. Note that the estimated state (blue in Fig. 2.14) is used to control the vehicle, thus the estimation errors are reflected in the plot of the ground-truth trajectory (gray in Fig. 2.14). Although the presented approaches are sufficient to drive autonomously the platform during some minutes without big trajectory errors, as stated before, the x and y positions and *yaw* angle are not observable (i.e., the method is an odometer) and their output is the result of an incremental estimation subject to drift.

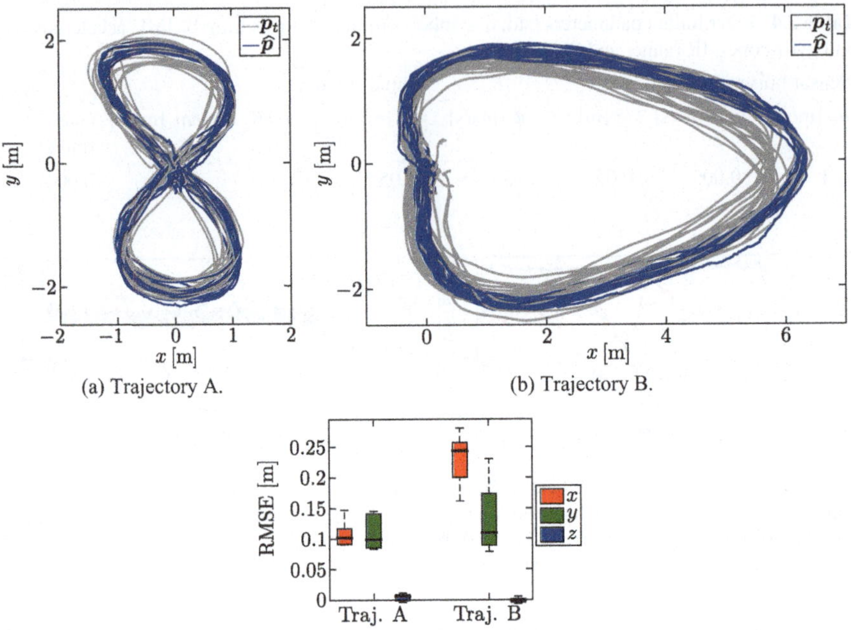

(a) Trajectory A.

(b) Trajectory B.

(c) Positioning RMSE statistics.

Fig. 2.13 Error analysis of two trajectories with 25 runs each. All runs are executed fully autonomously with a maximum cruise velocity of 1 m/s (best seen in color)

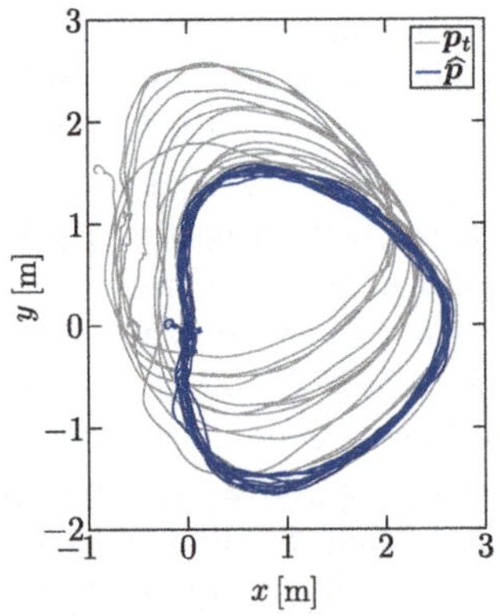

Fig. 2.14 Position estimation results for a long experiment (almost 10 min of continuous flight and a full battery discharge). Note that in full autonomous mode the vehicle is controlled using the estimation, thus the drift of the platform can be seen in the ground-truth trajectory (Qualisys motion capture system)

2.11 Conclusions

In this chapter, we presented a state estimator design for MAVs that combines low-cost and high-rate visual-inertial-range sensors. We investigated a wide range of algorithm variations with different computing and implementation complexities. We have shown the feasibility of using such low-cost sensor setup with light algorithms to achieve not only hovering maneuvers but also fully autonomous navigation. This research work has been partially published in [38] and [37]. All the technical details have been provided, facilitating the use of the proposed methods by other groups in the community.

The result of our experimentation shows that the effects of all the variations in the estimator design are minimal. In particular, the refinements on the transition matrices $\boldsymbol{F}_1 \cdots \boldsymbol{F}_3$ have no conclusive effect, meaning that the classical Euler approximation $\boldsymbol{F}_1$ is sufficiently good. A similar conclusion can be drawn for the quaternion integrators Q0B, Q0F and Q1, and even for the error compositions LE and GE. We conclude that the final choices can be driven more by a criterion of convenience rather than performance. This is due to the high frequency of the measurements and filter updates, which renders all integration schemes close to the continuous-time case, and therefore equivalent in practice. Regarding the filter type, EKF vs. ESKF, we also found equivalent performances. We can base our choice on different criteria. For example, EKF is more widely known, and it is also simpler, both conceptually and in terms of implementation complexity. However, ESKF is very close to it, and constitutes a more proper and elegant solution, from a theoretical viewpoint, because of its operation in the rotations manifold $\mathscr{SO}(3)$. This implies, for example, that in ESKF there is no need to perform quaternion re-normalization. Our recommendations are the classical EKF with $\boldsymbol{F}_1$, Q0B and quaternion re-normalization; or the more proper ESKF with $\boldsymbol{F}_1$, Q0B, and either GE or LE. Both have essentially the same computational cost.

Using these filters, in terms of overall precision, our state estimates are usable during flight times of several minutes, enabling the MAV to perform a number of tasks that require navigation without the aid on any external positioning system. With a MAV as a UAM aerial platform, these localization and control modules provide autonomy to the robot during navigation phases where precise manipulation is not required (e.g., approaching phase where the target is not yet on sight).

The estimated state is richer than just odometry, and includes higher derivatives such as velocities and accelerations, all precisely referenced to the gravity direction. These are exploited by a non-linear controller to drive the vehicle in 3D space, showing that the employed sensors are more than sufficient to provide autonomy to an aerial platform. This is the first time that such inexpensive sensors enable precise localization and autonomous navigation of aerial vehicles.

Appendix 2.A Error State Kinematics (Continous Time)

To obtain the error-state kinematic equations we have to write each state equation in (2.28) as its composition of nominal- and error-state, solve for the error-state and simplify all second-order infinitesimals. For the terms $\delta\dot{\boldsymbol{p}}$, $\delta\dot{\boldsymbol{a}}_b$, $\delta\dot{\boldsymbol{\omega}}_b$ this operation is trivial. In contrast, for the terms involving rotation elements ($\delta\dot{\boldsymbol{v}}$ and $\delta\dot{\boldsymbol{q}}$) require some non-trivial manipulations which are detailed in the following depending the orientation error definition (GE or LE).

2.A.1 Globally-Defined Orientation Error (GE)

2.A.1.1 GE: Linear Velocity Error

Considering the true acceleration as large- and small-signal accelerations in body frame $\boldsymbol{a}_{Bt} = \boldsymbol{a}_B + \delta\boldsymbol{a}_B$ with

$$\boldsymbol{a}_B \triangleq \boldsymbol{a}_s - \boldsymbol{a}_b\,, \tag{2.93a}$$

$$\delta\boldsymbol{a}_B \triangleq -\delta\boldsymbol{a}_b - \boldsymbol{a}_n\,, \tag{2.93b}$$

and substituting in (2.28b) we have

$$\dot{\boldsymbol{v}}_t = \boldsymbol{R}_t\,(\boldsymbol{a}_B + \delta\boldsymbol{a}_B) + \boldsymbol{g}\,. \tag{2.94}$$

Defining $\boldsymbol{v}_t$ as the composition of the nominal- plus the error-state, and $\boldsymbol{R}_t$ with its small signal approximation (i.e., $\boldsymbol{R}_t = (\boldsymbol{I} + [\delta\boldsymbol{\theta}]_\times)\boldsymbol{R} + \boldsymbol{O}(||\delta\boldsymbol{\theta}||^2)$) ignoring the infinitesimal term $\boldsymbol{O}(||\delta\boldsymbol{\theta}||^2)$, we end up with

$$\dot{\boldsymbol{v}} + \delta\dot{\boldsymbol{v}} = (\boldsymbol{I} + [\delta\boldsymbol{\theta}]_\times)\boldsymbol{R}\,(\boldsymbol{a}_B + \delta\boldsymbol{a}_B) + \boldsymbol{g}\,. \tag{2.95}$$

Substituting the nominal velocity $\dot{\boldsymbol{v}}$ by (2.30b), then using (2.93a) and finally rearranging terms, we have

$$\delta\dot{\boldsymbol{v}} = \boldsymbol{R}\delta\boldsymbol{a}_B + [\delta\boldsymbol{\theta}]_\times\boldsymbol{R}(\boldsymbol{a}_B + \delta\boldsymbol{a}_B)\,. \tag{2.96}$$

Reorganizing some cross-products (with $[\boldsymbol{a}]_\times\boldsymbol{b} = -[\boldsymbol{b}]_\times\boldsymbol{a}$), we get

$$\delta\dot{\boldsymbol{v}} = \boldsymbol{R}\delta\boldsymbol{a}_B - [\boldsymbol{R}\boldsymbol{a}_B]_\times\delta\boldsymbol{\theta}\,, \tag{2.97}$$

which recalling (2.93a) and (2.93b), and rearranging leads to

$$\delta\dot{\boldsymbol{v}} = -[\boldsymbol{R}(\boldsymbol{a}_s - \boldsymbol{a}_b)]_\times\delta\boldsymbol{\theta} - \boldsymbol{R}\delta\boldsymbol{a}_b - \boldsymbol{R}\boldsymbol{a}_n\,. \tag{2.98}$$

Assuming accelerometers noise as white, uncorrelated and isotropic, we obtain the dynamics of the linear velocity error

$$\delta\dot{\boldsymbol{v}} = -[\boldsymbol{R}(\boldsymbol{a}_s - \boldsymbol{a}_b)]_\times\delta\boldsymbol{\theta} - \boldsymbol{R}\delta\boldsymbol{a}_b - \boldsymbol{a}_n\,. \tag{2.99}$$

2.A.1.2 GE: Orientation Error

With the true quaternion $\boldsymbol{q}_t$ as a composition of the nominal- and error-state rotations, and its derivation from (2.10), we have

$$\begin{aligned}
\dot{\boldsymbol{q}}_t &= (\delta \boldsymbol{q} \overset{\cdot}{\otimes} \boldsymbol{q}) && (2.100a)\\
&= \delta \dot{\boldsymbol{q}} \otimes \boldsymbol{q} + \delta \boldsymbol{q} \otimes \dot{\boldsymbol{q}} && (2.100b)\\
&= \delta \dot{\boldsymbol{q}} \otimes \boldsymbol{q} + \frac{1}{2}\delta \boldsymbol{q} \otimes \boldsymbol{q} \otimes \boldsymbol{\omega}\,. && (2.100c)
\end{aligned}$$

Similarly, we can define

$$\begin{aligned}
\dot{\boldsymbol{q}}_t &= \frac{1}{2}\boldsymbol{q}_t \otimes \boldsymbol{\omega}_t && (2.101a)\\
&= \frac{1}{2}\delta \boldsymbol{q} \otimes \boldsymbol{q} \otimes \boldsymbol{\omega}_t\,. && (2.101b)
\end{aligned}$$

Matching (2.100c) with (2.101b), and having $\boldsymbol{\omega}_t = \boldsymbol{\omega} + \delta\boldsymbol{\omega}$, this reduces to

$$\delta \dot{\boldsymbol{q}} \otimes \boldsymbol{q} = \frac{1}{2}\delta \boldsymbol{q} \otimes \boldsymbol{q} \otimes \delta\boldsymbol{\omega}\,. \tag{2.102}$$

Right-multiplying left and right terms by $\boldsymbol{q}^*$, and recalling that $\boldsymbol{q} \otimes \delta\boldsymbol{\omega} \otimes \boldsymbol{q}^* \equiv \boldsymbol{R}\delta\boldsymbol{\omega}$, we can further develop as follows

$$\begin{aligned}
\delta \dot{\boldsymbol{q}} &= \frac{1}{2}\delta \boldsymbol{q} \otimes \boldsymbol{q} \otimes \delta\boldsymbol{\omega} \otimes \boldsymbol{q}^* && (2.103a)\\
&= \frac{1}{2}\delta \boldsymbol{q} \otimes (\boldsymbol{R}\delta\boldsymbol{\omega}) && (2.103b)\\
&= \frac{1}{2}\delta \boldsymbol{q} \otimes \delta\boldsymbol{\omega}_G\,, && (2.103c)
\end{aligned}$$

with $\delta\boldsymbol{\omega}_G \equiv \boldsymbol{R}\delta\boldsymbol{\omega}$ the small-signal angular rate expressed in the global frame. Then,

$$\begin{aligned}
2\delta \dot{\boldsymbol{q}} &= \delta \boldsymbol{q} \otimes \delta\boldsymbol{\omega}_G && (2.104a)\\
&= \boldsymbol{\Omega}(\delta\boldsymbol{\omega}_G)\delta\boldsymbol{q} && (2.104b)\\
&= \begin{bmatrix} 0 & -\delta\boldsymbol{\omega}_G^\top \\ \delta\boldsymbol{\omega}_G & -[\delta\boldsymbol{\omega}_G]_\times \end{bmatrix} \begin{bmatrix} 1 \\ \delta\boldsymbol{\theta}/2 \end{bmatrix}. && (2.104c)
\end{aligned}$$

Discarding the first row, which is not very useful, we can extract

$$\delta\dot{\boldsymbol{\theta}} = \delta\boldsymbol{\omega}_G - \frac{1}{2}[\delta\boldsymbol{\omega}_G]_\times \delta\boldsymbol{\theta}\,, \tag{2.105}$$

where also removing the second order infinitesimal terms

$$\delta\dot{\boldsymbol{\theta}} = \delta\boldsymbol{\omega} = \boldsymbol{R}\delta\boldsymbol{\omega}\,, \tag{2.106}$$

where recalling (2.119b) we obtain the linearized dynamics of the global angular error,

$$\delta\dot{\boldsymbol{\theta}} = -\boldsymbol{R}\delta\boldsymbol{\omega}_b - \boldsymbol{R}\delta\boldsymbol{\omega}_n \,. \tag{2.107}$$

2.A.2 *Locally-Defined Orientation Error (LE)*

2.A.2.1 LE: Linear Velocity Error

Starting from (2.94) and defining $\boldsymbol{v}_t$ as the composition of the nominal- plus the error-state, and $\boldsymbol{R}_t$ with its small signal approximation (i.e., $\boldsymbol{R}_t = \boldsymbol{R}\,(\boldsymbol{I} + [\delta\boldsymbol{\theta}]_\times) + \boldsymbol{O}(||\delta\boldsymbol{\theta}||^2)$) ignoring the infinitesimal term $\boldsymbol{O}(||\delta\boldsymbol{\theta}||^2)$, we end up with

$$\dot{\boldsymbol{v}} + \delta\dot{\boldsymbol{v}} = \boldsymbol{R}\,(\boldsymbol{I} + [\delta\boldsymbol{\theta}]_\times)\,(\boldsymbol{a}_B + \delta\boldsymbol{a}_B) + \boldsymbol{g} \,. \tag{2.108}$$

Substituting the nominal velocity $\dot{\boldsymbol{v}}$ by (2.30b), then using (2.93a) and finally rearranging terms, we have

$$\delta\dot{\boldsymbol{v}} = \boldsymbol{R}[\delta\boldsymbol{\theta}]_\times\boldsymbol{a}_B + \boldsymbol{R}\delta\boldsymbol{a}_B + \boldsymbol{R}[\delta\boldsymbol{\theta}]_\times\,\delta\boldsymbol{a}_B \,. \tag{2.109}$$

Reorganizing some cross-products (with $[\boldsymbol{a}]_\times\boldsymbol{b} = -[\boldsymbol{b}]_\times\boldsymbol{a}$), we get

$$\delta\dot{\boldsymbol{v}} = \boldsymbol{R}(\delta\boldsymbol{a}_B - [\boldsymbol{a}_B]_\times\delta\boldsymbol{\theta}) \,, \tag{2.110}$$

which recalling (2.93a) and (2.93b), and rearranging leads to

$$\delta\dot{\boldsymbol{v}} = -\boldsymbol{R}[\boldsymbol{a}_s - \boldsymbol{a}_b]_\times\delta\boldsymbol{\theta} - \boldsymbol{R}\delta\boldsymbol{a}_b - \boldsymbol{R}\boldsymbol{a}_n \,. \tag{2.111}$$

Assuming accelerometers noise as white, uncorrelated and isotropic, we obtain the dynamics of the linear velocity error

$$\delta\dot{\boldsymbol{v}} = -\boldsymbol{R}[\boldsymbol{a}_s - \boldsymbol{a}_b]_\times\delta\boldsymbol{\theta} - \boldsymbol{R}\delta\boldsymbol{a}_b - \boldsymbol{a}_n \,. \tag{2.112}$$

2.A.2.2 LE: Orientation Error

With the true quaternion $\boldsymbol{q}_t$ as a composition of the nominal- and error-state rotations, and the quaternion derivative from (2.10), we have

$$\dot{\boldsymbol{q}}_t = (\boldsymbol{q} \overset{\bullet}{\otimes} \delta\boldsymbol{q}) \tag{2.113a}$$

$$= \dot{\boldsymbol{q}} \otimes \delta\boldsymbol{q} + \boldsymbol{q} \otimes \delta\dot{\boldsymbol{q}} \tag{2.113b}$$

$$= \frac{1}{2}\boldsymbol{q} \otimes \boldsymbol{\omega} \otimes \delta\boldsymbol{q} + \boldsymbol{q} \otimes \delta\dot{\boldsymbol{q}} \,. \tag{2.113c}$$

Similarly, we can define

$$\dot{\boldsymbol{q}}_t = \frac{1}{2}\boldsymbol{q}_t \otimes \boldsymbol{\omega}_t \tag{2.114a}$$

$$= \frac{1}{2}\boldsymbol{q} \otimes \delta\boldsymbol{q} \otimes \boldsymbol{\omega}_t\,. \tag{2.114b}$$

Matching (2.113c) with (2.114b), simplifying the leading $\boldsymbol{q}$ and isolating $\delta\dot{\boldsymbol{q}}$, we get

$$2\delta\dot{\boldsymbol{q}} = \delta\boldsymbol{q} \otimes \boldsymbol{\omega}_t - \boldsymbol{\omega} \otimes \delta\boldsymbol{q}\,, \tag{2.115}$$

where substituting (2.4) and (2.7c), we end up with

$$2\delta\dot{\boldsymbol{q}} = {}^{-}\boldsymbol{Q}(\boldsymbol{\omega}_t)\delta\boldsymbol{q} - {}^{+}\boldsymbol{Q}(\boldsymbol{\omega})\delta\boldsymbol{q} \tag{2.116a}$$

$$= \left({}^{-}\boldsymbol{Q}(\boldsymbol{\omega}_t) - {}^{+}\boldsymbol{Q}(\boldsymbol{\omega})\right)\begin{bmatrix} 1 \\ \delta\boldsymbol{\theta}/2 \end{bmatrix} \tag{2.116b}$$

$$= \begin{bmatrix} 0 & -(\boldsymbol{\omega}_t - \boldsymbol{\omega})^\top \\ (\boldsymbol{\omega}_t - \boldsymbol{\omega}) & -[\boldsymbol{\omega}_t + \boldsymbol{\omega}]_\times \end{bmatrix}\begin{bmatrix} 1 \\ \delta\boldsymbol{\theta}/2 \end{bmatrix} \tag{2.116c}$$

$$= \begin{bmatrix} 0 & -\delta\boldsymbol{\omega}^\top \\ \delta\boldsymbol{\omega} & -[2\boldsymbol{\omega} + \delta\boldsymbol{\omega}]_\times \end{bmatrix}\begin{bmatrix} 1 \\ \delta\boldsymbol{\theta}/2 \end{bmatrix}\,. \tag{2.116d}$$

Simplifying again with (2.7c) the derivative term, we obtain

$$\begin{bmatrix} 0 \\ \delta\dot{\boldsymbol{\theta}} \end{bmatrix} = \begin{bmatrix} 0 & -\delta\boldsymbol{\omega}^\top \\ \delta\boldsymbol{\omega} & -[2\boldsymbol{\omega} + \delta\boldsymbol{\omega}]_\times \end{bmatrix}\begin{bmatrix} 1 \\ \delta\boldsymbol{\theta}/2 \end{bmatrix}\,. \tag{2.117}$$

Discarding the first row, which is not very useful, we can extract

$$\delta\dot{\boldsymbol{\theta}} = \delta\boldsymbol{\omega} - [\boldsymbol{\omega}]_\times\delta\boldsymbol{\theta} - \frac{1}{2}[\delta\boldsymbol{\omega}]_\times\delta\boldsymbol{\theta}\,, \tag{2.118}$$

where also removing the second order infinitesimal terms and considering the true angular velocities as a composition of its large- and small-signal elements $\boldsymbol{\omega}_t = \boldsymbol{\omega} + \delta\boldsymbol{\omega}$ with

$$\boldsymbol{\omega} \triangleq \boldsymbol{\omega}_s - \boldsymbol{\omega}_b\,, \tag{2.119a}$$

$$\delta\boldsymbol{\omega} \triangleq -\delta\boldsymbol{\omega}_b - \boldsymbol{\omega}_n\,, \tag{2.119b}$$

we obtain the dynamics of the orientation error

$$\delta\dot{\boldsymbol{\theta}} = -[\boldsymbol{\omega}_s - \boldsymbol{\omega}_b]_\times\delta\boldsymbol{\theta} - \delta\boldsymbol{\omega}_b - \delta\boldsymbol{\omega}_n\,. \tag{2.120}$$

Appendix 2.B Rotation Matrix Partial Derivatives

In order to differentiate the rotation of a vector, $\boldsymbol{q}\boldsymbol{\vartheta}\boldsymbol{q}^*$, it is convenient to use the matrix form $\boldsymbol{R}(\boldsymbol{q})$. Herein, we compute an example of the derivatives of the function

$$h(\boldsymbol{x}) = \boldsymbol{R}^\top \boldsymbol{\vartheta} + \boldsymbol{n}\,, \tag{2.121}$$

with $\boldsymbol{n}$ the measurement noise assumed to be Gaussian with zero mean and covariance matrix $\boldsymbol{n}$. The corresponding derivatives are detailed in the following sections depending on whether they are computed with respect to the quaternion value (used in EKF) or with the orientation error with minimal representation (used in ESKF).

2.B.1 Partial Derivative w.r.t. Quaternion

To compute $\partial h(\boldsymbol{x})/\partial \boldsymbol{q}$ (the subscripts t is avoided here for clarity) is more convenient to express the derivatives w.r.t. each component i of the quaternion $\boldsymbol{q}$ such as

$$c\partial \left(\boldsymbol{R}^\top \boldsymbol{\vartheta}\right)\partial \boldsymbol{q}_i = \boldsymbol{R}^\top \frac{\partial \boldsymbol{\vartheta}}{\partial \boldsymbol{q}_i} + \frac{\partial \boldsymbol{R}^\top}{\partial \boldsymbol{q}_i}\boldsymbol{\vartheta}\,. \tag{2.122}$$

With the definition of (2.9) we get

$$c\partial \left(\boldsymbol{R}^\top \boldsymbol{\vartheta}\right)\partial \boldsymbol{q}_w = 2\begin{bmatrix} q_w & q_z & -q_y \\ -q_z & q_w & q_x \\ q_y & -q_x & q_w \end{bmatrix}\boldsymbol{\vartheta}\,, \tag{2.123a}$$

$$c\partial \left(\boldsymbol{R}^\top \boldsymbol{\vartheta}\right)\partial \boldsymbol{q}_x = 2\begin{bmatrix} q_x & -q_y & -q_z \\ -q_y & -q_x & q_w \\ -q_z & -q_w & -q_x \end{bmatrix}\boldsymbol{\vartheta}\,, \tag{2.123b}$$

$$c\partial \left(\boldsymbol{R}^\top \boldsymbol{\vartheta}\right)\partial \boldsymbol{q}_y = 2\begin{bmatrix} -q_y & -q_x & -q_w \\ -q_x & q_y & -q_z \\ q_w & -q_z & -q_y \end{bmatrix}\boldsymbol{\vartheta}\,, \tag{2.123c}$$

$$c\partial \left(\boldsymbol{R}^\top \boldsymbol{\vartheta}\right)\partial \boldsymbol{q}_z = 2\begin{bmatrix} -q_z & q_w & -q_x \\ -q_w & -q_z & -q_y \\ -q_x & -q_y & q_z \end{bmatrix}\boldsymbol{\vartheta}\,. \tag{2.123d}$$

2.B.2 Partial Derivative w.r.t. Rotation Error Using Minimal Representation

We now want to compute $\partial h(\boldsymbol{x}_t)/\partial \delta\boldsymbol{\theta}$, which will depend on the orientation error representation (GE or LE) as explained in the following.

2.B.2.1 Globally-Defined Orientation Error (GE)

If we define the orientation error globally (i.e. $\boldsymbol{q}_t = \delta\boldsymbol{q} \otimes \boldsymbol{q}$) and with minimal representation $\delta\boldsymbol{\theta}$ as shown in (2.7c), we have

$$\begin{aligned} fx_t) &= \boldsymbol{R}_t(\boldsymbol{\theta}_t)^\top \boldsymbol{\vartheta} \\ & \boldsymbol{R}(\boldsymbol{\theta})^\top \boldsymbol{R}(\delta\boldsymbol{\theta})^\top \boldsymbol{\vartheta} \,. \end{aligned} \tag{2.124}$$

Taking advantage of the rotation matrix approximation $\boldsymbol{R}(\delta\boldsymbol{\theta})^\top \approx (\boldsymbol{I} - [\delta\boldsymbol{\theta}]_\times)$, we end up with

$$\begin{aligned} h(\boldsymbol{x}_t) &\approx \boldsymbol{R}_t(\boldsymbol{\theta})^\top (\boldsymbol{I} - [\delta\boldsymbol{\theta}]_\times)\boldsymbol{\vartheta} \\ &\approx \boldsymbol{R}(\boldsymbol{\theta})^\top \boldsymbol{\vartheta} - \boldsymbol{R}(\boldsymbol{\theta})^\top [\delta\boldsymbol{\theta}]_\times \boldsymbol{\vartheta} \\ &\approx \boldsymbol{R}(\boldsymbol{\theta})^\top \boldsymbol{\vartheta} + \boldsymbol{R}(\boldsymbol{\theta})^\top [\boldsymbol{\vartheta}]_\times \delta\boldsymbol{\theta} \,. \end{aligned} \tag{2.125}$$

Thus, the derivative with respect to the orientation error is

$$\frac{\partial h(\boldsymbol{x}_t)}{\partial \delta\boldsymbol{\theta}} = \boldsymbol{R}(\boldsymbol{\theta})^\top [\boldsymbol{\vartheta}]_\times \,. \tag{2.126}$$

2.B.2.2 Locally-Defined Orientation Error (LE)

Similarly to the previous derivation, if the orientation error is defined locally (i.e. $\boldsymbol{q}_t = \boldsymbol{q} \otimes \delta\boldsymbol{q}$) and with minimal representation, we have

$$\begin{aligned} h(\boldsymbol{x}_t) &= \boldsymbol{R}_t(\boldsymbol{\theta}_t)^\top \boldsymbol{\vartheta} \\ &= \boldsymbol{R}(\delta\boldsymbol{\theta})^\top \boldsymbol{R}(\boldsymbol{\theta})^\top \boldsymbol{\vartheta} \,. \end{aligned} \tag{2.127}$$

Using the rotation matrix approximation $\boldsymbol{R}(\delta\boldsymbol{\theta})^\top \approx (\boldsymbol{I} - [\delta\boldsymbol{\theta}]_\times)$, we end up with

$$\begin{aligned} h(\boldsymbol{x}_t) &\approx (\boldsymbol{I} - [\delta\boldsymbol{\theta}]_\times)\boldsymbol{R}(\boldsymbol{\theta})^\top \boldsymbol{\vartheta} \\ &\approx \boldsymbol{R}(\boldsymbol{\theta})^\top \boldsymbol{\vartheta} - [\delta\boldsymbol{\theta}]_\times \boldsymbol{R}_t(\boldsymbol{\theta})^\top \boldsymbol{\vartheta} \\ &\approx \boldsymbol{R}(\boldsymbol{\theta})^\top \boldsymbol{\vartheta} + [\boldsymbol{R}(\boldsymbol{\theta})^\top \boldsymbol{\vartheta}]_\times \delta\boldsymbol{\theta}\,. \end{aligned} \tag{2.128}$$

So, the derivative with respect to the orientation error is

$$\frac{\partial h(\boldsymbol{x}_t)}{\partial \delta\boldsymbol{\theta}} = [\boldsymbol{R}(\boldsymbol{\theta})^\top \boldsymbol{\vartheta}]_\times\,. \tag{2.129}$$

References

1. Bar-Shalom, Yaakov, X. Rong Li, and Thiagalingam Kirubarajan. 2004. *Estimation with Applications to Tracking and Navigation: Theory Algorithms and Software*. Wiley.
2. Blösch, Michael, Sammy Omari, Péter Fankhauser, Hannes Sommer, Christian Gehring, Jemin Hwangbo, Mark A. Hoepflinger, Marcus Hutter, and Roland Siegwart. 2014. Fusion of optical flow and inertial measurements for robust egomotion estimation. In *IEEE/RSJ International Conference on Intelligent Robots and Systems*, 3102–3107. Chicago, USA, September 2014.
3. Blösch, Michael, Stephan Weiss, Davide Scaramuzza, and Roland Siegwart. Vision based MAV navigation in unknown and unstructured environments. In *IEEE International Conference on Robotics and Automation*, 21–28. Anchorage, USA, May 2010.
4. Bullo, Francesco, and Andrew D. Lewis. 2004. *Geometric Control of Mechanical Systems. Texts in Applied Mathematics*, vol. 49. Springer.
5. Engel, Jakob, Jürgen Sturm, and Daniel Cremers. 2012. Camera-based navigation of a low-cost quadrocopter. In *IEEE/RSJ International Conference on Intelligent Robots and Systems*, 2815–2821. Vilamoura, Portugal, October 2012.
6. Fliess, Michael, Jean Lévine, Philippe Martin, and Pierre Rouchon. 1995. Flatness and defect of non-linear systems: introductory theory and examples. *International Journal of Control* 61 (6): 1327–1361.
7. Forte, Francesco, Roberto Naldi, Alessandro Macchelli, and Lorenzo Marconi. 2012. Impedance control of an aerial manipulator. In *American Control Conference*, 3839–3844. Montréal, Canada, June 2012.
8. Fraundorfer, Friedrich, Lionel Heng, Dominik Honegger, Gim Hee Lee, Lorenz Meier, Petri Tanskanen, and Marc Pollefeys. 2012. Vision-based autonomous mapping and exploration using a quadrotor MAV. In *IEEE/RSJ International Conference on Intelligent Robots and Systems*, 4557–4564. Vilamoura, Portugal, October 2012.
9. Hesch, Joel A., Dimitrios G. Kottas, Sean L. Bowman, and Stergios I. Roumeliotis. 2013. Camera-IMU-based localization: Observability analysis and consistency improvement. *The International Journal of Robotics Research* 33 (1): 182–201.
10. Honegger, Dominik, Lorenz Meier, Petri Tanskanen, and Marc Pollefeys. 2013. An open source and open hardware embedded metric optical flow cmos camera for indoor and outdoor applications. In *IEEE International Conference on Robotics and Automation*, 1736–1741. Karlsruhe, Germany, May 2013.
11. Jones, Eagle S., and Stefano Soatto. 2011. Visual-inertial navigation, mapping and localization: A scalable real-time causal approach. *The International Journal of Robotics Research* 30 (4): 407–430.
12. Kelly, Jonathan, and Gaurav S. Sukhatme. 2007. An experimental study of aerial stereo visual odometry. *IFAC Proceedings Volumes* 40 (15): 197–202.

13. Kelly, Jonathan, and Gaurav S. Sukhatme. 2011. Visual-inertial sensor fusion: Localization, mapping and sensor-to-sensor self-calibration. *The International Journal of Robotics Research* 30 (1): 56–79.
14. Konolige, Kurt, Motilal Agrawal, and Joan Sola. 2011. Large-Scale Visual Odometry for Rough Terrain. *Robotics Research: The 13th International Symposium ISRR*, vol. 66, 201–212. Berlin Heidelberg: Springer.
15. Lee, Taeyoung, Melvin Leok, and N. Harris McClamroch. 2013. Nonlinear robust tracking control of a quadrotor UAV on SE (3). *Asian Journal of Control*, 15 (2): 391–408.
16. Li, Mingyang, and Anastasios I. Mourikis. 2012. Improving the accuracy of EKF-based visual-inertial odometry. In *IEEE International Conference on Robotics and Automation*, 828–835. St. Paul, USA, May 2012.
17. Li, Mingyang, and Anastasios I. Mourikis. 2013. High-precision, consistent EKF-based visual-inertial odometry. *The International Journal of Robotics Research* 32 (6): 690–711.
18. Lim, Hyon, Sudipta N. Sinha, Michael F. Cohen, and Matthew Uyttendaele. 2012. Real-time image-based 6-dof localization in large-scale environments. In *IEEE Conference on Computer Vision and Pattern Recognition*, 1043–1050. Providence, USA, June 2012.
19. Liu, Hui, Houshang Darabi, Pat Banerjee, and Jing Liu. 2007. Survey of wireless indoor positioning techniques and systems. *IEEE Transactions on Systems, Man, and Cybernetics* 37 (6): 1067–1080.
20. Loianno, G., Y. Mulgaonkar, C. Brunner, D. Ahuja, A. Ramanandan, M. Chari, S. Diaz, and V. Kumar. 2015. Smartphones power flying robots. In *IEEE/RSJ International Conference on Intelligent Robots and Systems*, 1256–1263. Hamburg, Germany, September 2015.
21. Loianno, Giuseppe, Justin Thomas, and Vijay Kumar. 2015. Cooperative localization and mapping of MAVs using RGB-D sensors. In *IEEE International Conference on Robotics and Automation*, 4021–4028. Seattle, USA, May 2015.
22. Lupton, Todd, and Salah Sukkarieh. 2012. Visual-inertial-aided navigation for high-dynamic motion in built environments without initial conditions. *IEEE Transactions on Robotics* 28 (1): 61–76.
23. Madyastha, Venkatesh, Vishal Ravindra, Srinath Mallikarjunan, and Anup Goyal. 2011. Extended Kalman filter vs. error state Kalman filter for aircraft attitude estimation. *AIAA Guidance*, 6615–6638. Navigation, and Control Conference Oregon: Portland, August 2011.
24. Martinelli, Agostino. 2012. Vision and IMU data fusion: Closed-form solutions for attitude, speed, absolute scale, and bias determination. *IEEE Transactions on Robotics* 28 (1): 44–60.
25. Martinelli, Alessio. 2013. Visual-inertial structure from motion: observability and resolvability. In *IEEE/RSJ International Conference on Intelligent Robots and Systems*, 4235–4242. Tokyo, Japan, November 2013.
26. Mellinger, Daniel, and Vijay Kumar. 2011. Minimum snap trajectory generation and control for quadrotors. In *IEEE International Conference on Robotics and Automation*, 2520–2525. Shanghai, China, May 2011.
27. Michael, Nathan, Daniel Mellinger, Quentin Lindsey, and Vijay Kumar. 2010. The GRASP multiple micro-UAV testbed. *IEEE Robotics & Automation Magazine* 17 (3): 56–65.
28. Michael, Nathan, Shaojie Shen, Kartik Mohta, Yash Mulgaonkar, Vijay Kumar, Keiji Nagatani, Yoshito Okada, Seiga Kiribayashi, Kazuki Otake, and Kazuya Yoshida, Ohno, Kazunori, Takeuchi, Eijiro, and Tadokoro, Satoshi. 2012. Collaborative mapping of an earthquake-damaged building via ground and aerial robots. *Journal of Field Robotics* 29 (5): 832–841.
29. Mourikis, Anastasios I., and Stergios I. Roumeliotis. 2007. A multi-state constraint Kalman filter for vision-aided inertial navigation. In *IEEE International Conference on Robotics and Automation*, 3565–3572. Roma, Italy, April 2007.
30. Nikolic, Janosch, Joern Rehder, Michael Burri, Pascal Gohl, Stefan Leutenegger, Paul T. Furgale, and Roland Siegwart. A synchronized visual-inertial sensor system with FPGA preprocessing for accurate real-time SLAM. In *IEEE International Robotics and Automation*, 431–437. Hong Kong, China, May 2014.
31. Omari, S., and G. Ducard. 2013. Metric visual-inertial navigation system using single optical flow feature. In *European Control Conference*, 1310–1316. Aalborg, Denmark, July 2013.

32. Ravindra, Vishal C., Venkatesh K. Madyastha, and Anup Goyal. 2012. The equivalence between two well-known variants of the Kalman filter. In *International Conference on Advances in Control and Optimization of Dynamic Systems*. Bangalore, India, February 2012.
33. Roumeliotis, Stergios I., Andrew E. Johnson, and James F. Montgomery. 2002. Augmenting inertial navigation with image-based motion estimation. In *IEEE International Conference on Robotics and Automation*, vol. 4, 4326–4333. Washington, USA, May 2002.
34. Roussillon, Cyril, Aurélien Gonzalez, Joan Solá, Jean-Marie Codol, Nicolas Mansard, Simon Lacroix, and Michel Devy. 2011. RT-SLAM: a Generic and Real-Time Visual SLAM Implementation. In *Computer Vision Systems. Lecture Notes in Computer Science*, vol. 6962, 31–40. Springer.
35. Ruffo, M., M. Di Castro, L. Molinari, R. Losito, A. Masi, J. Kovermann, and Luis Rodrigues. 2014. New infrared time of-flight measurement sensor for robotic platforms. In *20th IMEKO TC4 International Symposium and Workshop on ADC Modelling and Testing*. Benevento, Italy, September 2014.
36. Santamaria-Navarro, Angel, and Juan Andrade-Cetto. 2013. Uncalibrated image-based visual servoing. In *IEEE International Conference on Robotics and Automation*, 5247–5252. Karlsruhe, Germany, May 2013.
37. Santamaria-Navarro, Angel, Giuseppe Loianno, Joan Solá Vijay Kumar, and Juan Andrade-Cetto. 2017. Autonomous navigation of micro aerial vehicles: State estimation using fast and low-cost sensors. *Submitted to Autonomous Robots*.
38. Santamaria-Navarro, Angel, Joan Sola, and Juan Andrade-Cetto. 2015. High-frequency MAV state estimation using low-cost inertial and optical flow measurement units. In *IEEE/RSJ International Conference on Intelligent Robots and Systems*, 1864–1871. Hamburg, Germany, October 2015.
39. Shen, Shaojie, Nathan Michael, and Vijay Kumar. 2012. Autonomous indoor 3d exploration with a micro-aerial vehicle. In *IEEE International Conference on Robotics and Automation*, 9–15. St. Paul, USA, May 2012.
40. Shen, Shaojie, Nathan Michael, and Vijay Kumar. 2015. Tightly-coupled monocular visual-inertial fusion for autonomous flight of rotorcraft MAVs. In *IEEE International Conference on Robotics and Automation*, 5303–5310. Seattle, USA, May 2015.
41. Shen, Shaojie, Yash Mulgaonkar, Nathan Michael, and Vijay Kumar. 2013. Vision-based state estimation and trajectory control towards high-speed flight with a quadrotor. In *Robotics: Science and Systems*, vol. 1. Berlin, Germany, June 2013.
42. Solá, Joan. 2012. Quaternion kinematics for the error-state KF. *Laboratoire d'Analyse et d'Architecture des Systemes-Centre national de la recherche scientifique (LAAS-CNRS) technical report*.
43. Solá, Joan, Teresa Vidal-Calleja, Javier Civera, and José María Martínez Montiel. 2012. Impact of landmark parametrization on monocular EKF-SLAM with points and lines. *International Journal of Computer Vision*, 97 (3): 339–368.
44. Thomas, Julian, Giuseppe Loianno, Koushil Sreenath, and Vipin Kumar. 2014. Toward image based visual servoing for aerial grasping and perching. In *IEEE International Conference on Robotics and Automation*, 2113–2118. Hong Kong, China, June 2014.
45. Thompson, R. 1985. A note on restricted maximum likelihood estimation with an alternative outlier model. *Journal of the Royal Statistical Society. Series B (Methodological)* 47 (1): 53–55.
46. Tomić, Teodor, Korbinian Schmid, Philipp Lutz, Andreas Dömel, Michael Kassecker, Elmar Mair, Iris Lynne Grixa, Felix Ruess, Michael Suppa, and Darius Burschka. 2012. Toward a fully autonomous UAV: Research platform for indoor and outdoor urban search and rescue. *IEEE Robotics & Automation Magazine*, 19 (3): 46–56.
47. Trawny, Nikolas, and Stergios I. Roumeliotis. 2005. Indirect Kalman filter for 3d attitude estimation. *University of Minnesota, Department of Computer Science and Engineering, Technical Report*, 2:2005.
48. Valenti, Roberto G., Ivan Dryanovski, Carlos Jaramillo, Daniel Perea Strom, and Jizhong Xiao. 2014. Autonomous quadrotor flight using onboard RGB-D visual odometry. In *IEEE International Conference on Robotics and Automation*, 5233–5238. Hong Kong, China, June 2014.

49. Weiss, Stephan, Markus W. Achtelik, Simon Lynen, Margarita Chli, and Roland Siegwart. 2012. Real-time onboard visual-inertial state estimation and self-calibration of mavs in unknown environments. In *IEEE International Conference on Robotics and Automation*, 957–964. Minneapolis, USA, May 2012.
50. Weiss, Stephan, Davide Scaramuzza, and Roland Siegwart. 2011. Monocular-SLAM based navigation for autonomous micro helicopters in GPS-denied environments. *Journal of Field Robotics* 28 (6): 854–874.
51. Wendel, Andreas, Arnold Irschara, and Horst Bischof. Natural landmark-based monocular localization for MAVs. In *IEEE International Conference on Robotics and Automation*, 5792–5799. Shanghai, China, May 2011.
52. özaslan, Tolga, Shaojie Shen, Yash Mulgaonkar, Nathan Michael, and Vijay Kumar. Inspection of penstocks and featureless tunnel-like environments using micro UAVs. In *Field and Service Robotics*. 123–136. Springer.

Chapter 3
Visual Servo

3.1 Introduction

UAMs require more precise localization capabilities than traditional UAVs, specially during manipulation phases (i.e., mission stages where the vehicle operates close to the target). Physical interaction with the environment calls for positioning accuracy at the centimeter level, which is often difficult to achieve. For indoor UAMs, accurate localization is usually obtained from infrared multi-camera systems, like Optitrack[5] or Vicon.[1] However, these devices are not suited for outdoor environments and require external infrastructure, as in the case of using GPS, which may not be practical.

To get rid of the external infrastructure dependency it is preferable to embark all perception hardware onboard, thus not relying on any external setup. This hardware must allow for target detection (i.e., using exteroceptive sensors) in order to achieve the desired mission interaction. Moreover, to choose the correct sensor suite, we must considering UAM restrictions in terms of payload, size and power consumption. Similarly to Chap. 2, a reasonable choice is to drive the vehicle using information provided by onboard cameras.

Visual servo (VS), also known as vision-based robot control, is a technique which uses feedback information extracted from one or multiple cameras to control the motion of a robot. The study and application of VS techniques include from image processing and sensor fusion, to control theory.

Regarding the VS algorithm, vision-based robot control systems are usually classified in three groups [3, 11]:

- Pose-based visual servo (PBVS)
- Image-based visual servo (IBVS)
- Hybrid control systems (HVS)

In PBVS, the geometric model of the target is used in conjunction with image features to estimate the pose of the target with respect to the camera frame. The

---

[1] www.vicon.com.

A. Santamaria-Navarro et al., *Visual Guidance of Unmanned Aerial Manipulators*, Springer Tracts in Advanced Robotics 125,
https://doi.org/10.1007/978-3-319-96580-2_3

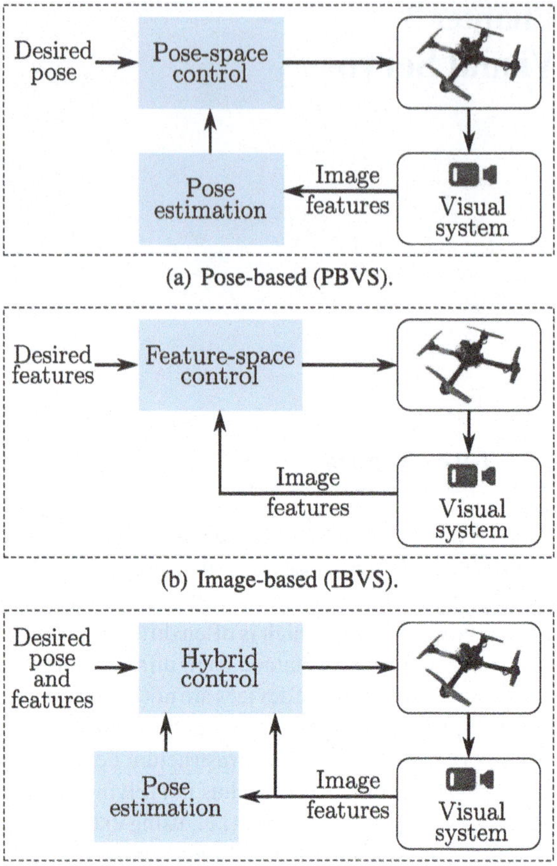

(a) Pose-based (PBVS).

(b) Image-based (IBVS).

(c) Hybrid (HVS).

Fig. 3.1 Visual servo schemes

control law is designed to reduce such pose error. However, it has the disadvantage that features could easily be lost in the image during the servo loop. In IBVS on the other hand, both the control objective and the control law are directly expressed in the image space, minimizing the error between observed and desired image feature coordinates. As a consequence, IBVS schemes do not need any a priori knowledge of the 3D structure of the observed scene. In addition, IBVS is more robust than PBVS with respect to uncertainties and disturbances affecting the model of the robot, as well as the calibration of the camera [10]. However, the convergence of IBVS methods is theoretically ensured only in a region around a desired position. To deal with PBVS and IBVS shortcomings, Hybrid methods, also called 2-1/2-D visual servo [16], combine IBVS and PBVS to estimate partial camera displacements at each iteration of the control law, minimizing a functional of both the pose error and the image feature error. These visual-control schemes are summarized in Fig. 3.1.

In this chapter we derive an uncalibrated image-based visual servo method (UIBVS) to drive the UAM. The proposed technique has the advantage that it contains mild assumptions about the principal point and skew values of the camera, and it does not require prior knowledge of the focal length, in contrast to traditional IBVS. Instead, the camera focal length is iteratively estimated within the control loop. Independence of focal length true value makes the system robust to noise and to unexpected large variations of this parameter (e.g., poor initialization or an unaccounted zoom change).

In this chapter we address the case where no constraints are placed on the camera motion, thus the system exhibits 6 DoF, and Sect. 3.7 shows its validations throughout simulation case studies. Further on, the UIBVS and HVS (which includes parts of PBVS) are also considered in Chap. 4 with real robot experiments.

This chapter is organized as follows. An overview of the state-of-art on visual servoing is given in the next section. For the sake of completeness, PBVS is briefly described in Sect. 3.4. IBVS details are presented in Sect. 3.5, and the UIBVS method is introduced in Sect. 3.6. Validation and simulations of the proposed method for UIBVS are provided in Sect. 3.7. Finally Sect. 3.8 concludes the chapter and provides a summary of main contributions.

3.2 Related Work

In 1980, [22] introduced a taxonomy of visual servo systems, into which the visual servo architectures can be categorized —see Fig. 3.1. Visual servoing concepts and notations were summarized afterward by [3, 4, 11].

The first group described, consists on PBVS approaches. [29] introduced a strategy to estimate the 3D camera pose but considering only planar motion in the visual control design. That work was extended in [30] to a unified approach for, the relative 3D pose estimation problem and, also for the full 3D visual servo control. A PBVS method using a nonlinear approach was presented in [17] by introducing 3D visual features in the closed robot control loop. A more recent work on PBVS is [14], where a suitable selection algorithm is adopted, at each sampling time, to allow the selection of an optimal set of visual data to be used for pose estimation. However, the dependency on 3D model information and well calibrated cameras makes this approach, and in general all PBVS methods, less atractive than image-based architectures.

The use of image-based approaches to control robots has been quite fruitful during the last decade. Some examples are [1, 5, 8]. Most of those methods follow the concepts introduced in [6], which sets the basis of error computation in the image plane and the formulation of the image Jacobian to translate those errors to Cartesian camera velocities. As an example, [18] proposes an IBVS together with a nonlinear observer for quadrotor velocity estimation, both using the same image information with the advantage that no extra sensor is required to measure translational velocities. Nevertheless, in most IBVS methods, error convergence to zero can typically be guaranteed only in a neighborhood of the desired configuration [2].

Error convergence to zero for the whole task can be obtained via a hybrid approach combining IBVS and PBVS, called 2-1/2-D visual servo, and introduced in [16]. This strategy estimates partial camera displacements at each iteration of the control law minimizing a functional of both, the error measures in image space typical from image-based servo and a log depth ratio accounting for the rate at which the camera moves to the target. [26] present a method to position the camera while keeping the target in its FoV with an IBVS approach. They first present kinematic modeling of the pose features with the objective of following a mobile target. Next, they define a control law adapted to ensure that the object remains in the camera FoV while achieving positioning tasks. A similar behavior is achieved in [13], but this time the PBVS and IBVS techniques are set in a hierarchical task composition and applied to UAM. In contrast, [28] presented a PBVS used to control the camera orientation, whereas an IBVS control law is used for the positioning. First, they construct an explicit solution for the rotational motion such that the translational kinematics is translated into equality constraints. Then, a convex optimization problem for the translational motion is formulated, where the visibility constraints are incorporated as inequality constraints.

In all image-based and hybrid approaches however, the resulting image Jacobian or interaction matrix, which relates the camera velocity to the image feature velocities, depends on a priori knowledge of the intrinsic camera parameters. To avoid that, [19] presents a model-free approach for the uncalibrated case. However, they use an additional path planner to introduce constraints in the desired trajectory in order to get a feasible path. As an alternative, the method presented in [27] to determine the motion and structure of a planar region can be used even if the camera parameters are unknown, however some knowledge about the scene must be available.

To do away with the depth dependence, one could optimize for the parameters in the image Jacobian whilst the error in the image plane is being minimized. This is done for instance, using Gauss-Newton to minimize the squared image error and non-linear least squares optimization for the image Jacobian [21]; using weighted recursive least squares (RLS), not to obtain the true parameters, but instead an approximation that still guarantees asymptotic stability of the control law in the sense of Lyapunov [9]; or using k-nearest neighbor regression to store previously estimated local models or previous movements, and estimating the Jacobian using local least squares (LLS) [7]. To provide robustness to outliers in the computation of the Jacobian, [25] proposes the use of an M-estimator.

We present a new approach to image-based visual servo in which the computation of the image Jacobian makes mild assumptions about the camera parameters. In particular, it assumes squared pixel size, centered principal point, and unknown focal length. Independence of focal length true value makes the system robust to noise and to unexpected large variations of this parameter (e.g., poor initialization or an unaccounted zoom change).

3.3 Control Law

The aim of all visual servo schemes is to drive the camera to a desired pose or point of view. In this section we present generic formulations to describe the basics of visual servoing control. Specific contents for PBVS, IBVS and UIBVS are provided in Sects. 3.4, 3.5 and 3.6 respectively. The description of hybrid approaches is not provided because it can be straightforwardly derived from PBVS and IBVS.

Let us consider $\boldsymbol{s}$ a generic variable, which can be a camera pose for PBVS or a set of features in the image plane for IBVS, depending on the visual servo approach used. In all visual servoing techniques we are interested in minimizing the error

$$\boldsymbol{e}(t) = \boldsymbol{s}(t) - \boldsymbol{s}^d, \tag{3.1}$$

where $\boldsymbol{s}(t)$ and $\boldsymbol{s}^d$ are the generic variables for its current and desired values respectively.

In pose-based methods $\boldsymbol{s}(t)$ corresponds to the camera pose. Computing this pose from a set of measurements in one image is a classical computer vision problem called the 3D localization problem, and requires the camera intrinsic parameters and the 3D model of the observed object to be known. Many solutions have been presented in the literature (e.g., [12]) to solve it, thus we consider this problem out of the scope of this chapter.

In image-based approaches $\boldsymbol{s}(t)$ are the current image coordinates of our set of target features, and $\boldsymbol{s}^d$ are their final desired position in the image plane.

To control the camera, the most straightforward approach is to design a velocity controller. To do this, we require the relationship between the time variation of the error $\boldsymbol{e}(t)$, and the camera velocity, which is defined as

$$\dot{\boldsymbol{e}} = \dot{\boldsymbol{s}} = \boldsymbol{J}\, {}^{c}\boldsymbol{\vartheta}, \tag{3.2}$$

with the camera velocity expressed in the current camera frame ${}^{c}\boldsymbol{\vartheta} = \left[{}^{c}\boldsymbol{v}^{\top}\ {}^{c}\boldsymbol{\omega}^{\top}\right]^{\top}$ and $\boldsymbol{J}$ the 6×6 so-called interaction Jacobian or image Jacobian in image-based methods. The Jacobians for each particular visual servo scheme are described in the following sections.

Assuming a holonomic system with 6 DoF, the camera velocities ${}^{c}\boldsymbol{\vartheta}$ can be used to command the robot with an exponential decoupled decrease of the error (i.e., $\dot{\boldsymbol{e}} = -\lambda \boldsymbol{e}$), such as

$${}^{c}\boldsymbol{\vartheta} = -\lambda\, \boldsymbol{J}^{+} \boldsymbol{e}, \tag{3.3}$$

where $\boldsymbol{J}^{+}$ is chosen as the Moore-Penrose pseudo-inverse of $\boldsymbol{J}$ (i.e., a 6×6 matrix), defined as

$$\boldsymbol{J}^{+} = (\boldsymbol{J}^{\top} \boldsymbol{J})^{-1} \boldsymbol{J}^{\top}. \tag{3.4}$$

This choice allows $||\dot{\boldsymbol{e}} - \lambda \boldsymbol{J} \boldsymbol{J}^{+} \boldsymbol{e}||$ and $||{}^{c}\boldsymbol{\vartheta}||$ to be minimal.

3.3.1 Stability Analysis

To analyze the stability of this closed loop visual servo system, we will use Lyapunov analysis. Let $\mathcal{L} = \frac{1}{2}\,\|\boldsymbol{e}(t)\|^2$ be a candidate Lyapunov function, whose derivative is given by [3]

$$\dot{\mathcal{L}} = \boldsymbol{e}^\top\,\dot{\boldsymbol{e}} \tag{3.5a}$$

$$= -\lambda\,\boldsymbol{e}^\top\,\boldsymbol{J}\,\boldsymbol{J}^+\,\boldsymbol{e}. \tag{3.5b}$$

Our system is globally asymptotic stable when the following sufficient condition holds

$$\boldsymbol{J}\,\boldsymbol{J}^+ > 0. \tag{3.6}$$

If the number of features used to computed $\boldsymbol{s}$ is, at least, equal to the number of camera DoF (i.e., 6), and if these features are chosen and the control scheme designed so that $\boldsymbol{J}$ and $\boldsymbol{J}^\top$ are of full rank 6, then condition (3.6) is ensured. With $\boldsymbol{J}$ non singular, we obtain from (3.6) the global asymptotic stability of the system since $\boldsymbol{J}\,\boldsymbol{J}^+ = \boldsymbol{I}$.

More details on stability analysis with particularizations for PBVS and IBVS can be found in many references (e.g., [3]).

3.4 Position-Based Visual Servo (PBVS)

In PBVS, also called 3D visual servo, the controller uses as input (i.e., the generic variable $\boldsymbol{s}$ in (3.1)) an estimation of the 3D pose of the camera with regard to a target object, obtained from an embedded camera and a reconstruction algorithm. PBVS computes the control error in 3D pose space and converts this error to camera velocities with the interaction Jacobian (3.2), defined in the following.

3.4.1 PBVS Interaction Jacobian

Considering the desired camera pose without movement (i.e., without linear or angular velocities), and following the frame definitions shown in Fig. 3.2, we can describe the current and desired poses, $\boldsymbol{s}$ and $\boldsymbol{s}^d$ respectively, and their error (3.1) with the pairs

$$\boldsymbol{s} = \{{}^{c^d}\boldsymbol{p}_c,\, {}^{c^d}\boldsymbol{R}_c\} \tag{3.7a}$$

$$\boldsymbol{s}^d = \{\boldsymbol{0},\, \boldsymbol{I}_3\} \tag{3.7b}$$

$$\boldsymbol{e} = \{\boldsymbol{e}_p,\, \boldsymbol{e}_\phi\} = \boldsymbol{s}, \tag{3.7c}$$

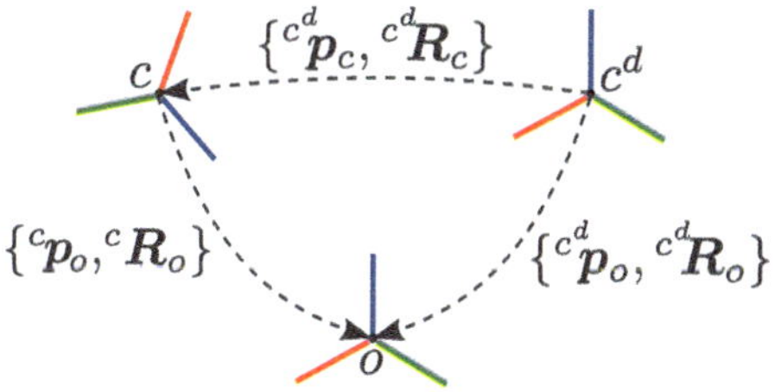

Fig. 3.2 Camera displacement for PBVS. The translation and orientation of a frame b expressed in a frame a is noted by the pair $\{{}^{a}\boldsymbol{p}_b, {}^{a}\boldsymbol{R}_b\}$. The involved coordinate frames are c and c^d corresponding to the current and desired camera poses, and o to the target object

where ${}^{c^d}\boldsymbol{p}_c$ and ${}^{c^d}\boldsymbol{R}_c$ are the camera position and its orientation respectively, both expressed in the desired camera frame c^d. Notice how the orientation error is here expressed as the rotation matrix ${}^{c^d}\boldsymbol{R}_c$, but it can also be represented by the corresponding Euler angles that form this matrix, the angle-axis representation (i.e., $\theta\boldsymbol{\varphi}$) or in quaternion form. In the latter, we can obtain the orientation error only computing the vector part of the relative quaternion rotation existing between frames, with

$$\boldsymbol{e}_\phi = {}^{c}\eta_o {}^{c^d}\boldsymbol{\epsilon}_o - {}^{c^d}\eta_o {}^{c}\boldsymbol{\epsilon}_o - \lfloor {}^{c^d}\boldsymbol{\epsilon}_o \rfloor_\times {}^{c}\boldsymbol{\epsilon}_o, \tag{3.8}$$

wehere $\{{}^{c}\eta_o, {}^{c}\boldsymbol{\epsilon}_o\}$ and $\{{}^{c^d}\eta_o, {}^{c^d}\boldsymbol{\epsilon}_o\}$ are the quaternions representing the orientations of frames c and c^d, respectively.

To obtain the derivatives of the error (3.7c), thus obtain the interaction Jacobian from (3.2), we have to solve separately for the translational ($\dot{\boldsymbol{e}}_p$) and angular ($\dot{\boldsymbol{e}}_\phi$) parts. We can generically describe the Jacobian with

$$\boldsymbol{J} = \begin{bmatrix} \boldsymbol{J}_p & \boldsymbol{0} \\ \boldsymbol{0} & \boldsymbol{J}_\phi \end{bmatrix}, \tag{3.9}$$

Notice how the decoupling between linear and angular parts is possible thanks to the error definition in (3.7c) (refer to [3] for more details).

To obtain the linear term we have to describe the velocity vector conversion from one frame to an other, thus having

$$\dot{\boldsymbol{e}}_p = {}^{c^d}\boldsymbol{v} = {}^{c^d}\boldsymbol{R}_c {}^{c}\boldsymbol{v}, \tag{3.10a}$$

where by simple inspection we obtain

$$\boldsymbol{J}_p = {}^{c^d}\boldsymbol{R}_c. \tag{3.11}$$

This operation can be defined using other rotation representations, but its development is here avoided for simplicity.

On the contrary, as the operations involving derivatives in $\mathscr{SO}(3)$ are not trivial, we show hereafter the expression of the angular error derivative depending on which representation do we choose (i.e., rotation matrices, quaternions or angle-axis).

Firstly, we use rotation matrices in the following way. From the *Rodrigues* formula, it can be shown that

$$\lfloor \boldsymbol{e}_\phi \rfloor_\times = \frac{1}{2}\left({}^{c^d}\boldsymbol{R}_c^\top - {}^{c^d}\boldsymbol{R}_c\right). \tag{3.12}$$

Deriving (3.12) together with the following relationship

$$\lfloor \dot{\boldsymbol{e}}_\phi \rfloor_\times = {}^{c^d}\dot{\boldsymbol{R}}_c^\top \, {}^{c^d}\boldsymbol{R}_c \tag{3.13a}$$

$$= -{}^{c^d}\boldsymbol{R}_c^\top \, {}^{c^d}\dot{\boldsymbol{R}}_c, \tag{3.13b}$$

we obtain

$$\lfloor \dot{\boldsymbol{e}}_\phi \rfloor_\times = \frac{1}{2}\left(\lfloor {}^c\boldsymbol{\omega} \rfloor_\times {}^{c^d}\boldsymbol{R}_c^\top + {}^{c^d}\boldsymbol{R}_c \lfloor {}^c\boldsymbol{\omega} \rfloor_\times\right), \tag{3.14}$$

where, after some computation (see [17]), we end up with

$$\dot{\boldsymbol{e}}_\phi = \frac{1}{2}\left(trace({}^{c^d}\boldsymbol{R}_c^\top)\,\boldsymbol{I} - {}^{c^d}\boldsymbol{R}_c\right) {}^c\boldsymbol{\omega}. \tag{3.15}$$

By simple inspection of (3.15) we obtain the orientation Jacobian using rotation matrices, corresponding to

$$\boldsymbol{J}_\phi = \frac{1}{2}\left(trace({}^{c^d}\boldsymbol{R}_c^\top)\,\boldsymbol{I} - {}^{c^d}\boldsymbol{R}_c\right). \tag{3.16}$$

Using quaternions, as in (3.8), we can take advantage of the relationship presented in (2.10), here summarized with

$$^c\dot{\boldsymbol{q}}_{c^d} = \frac{1}{2}\begin{bmatrix} 0 \\ {}^c\boldsymbol{\omega} \end{bmatrix} \otimes {}^c\boldsymbol{q}_{c^d}, \tag{3.17}$$

with $\otimes$ the quaternion product. Notice how the angular velocity is here composed globally in the camera frame (i.e., left quaternion product). Then, the derivative of the rotation part becomes

$$\dot{\boldsymbol{e}}_\phi = \frac{1}{2}\left({}^c\eta_{c^d}\boldsymbol{I}_3 - \lfloor {}^c\boldsymbol{\epsilon}_{c^d} \rfloor_\times\right) {}^c\boldsymbol{\omega}, \tag{3.18}$$

where by simple inspection we get the orientation Jacobian using quaternion expressions, defined as

$$\boldsymbol{J}_\phi = \frac{1}{2}\left({}^c\eta_{c^d}\boldsymbol{I}_3 - \lfloor {}^c\boldsymbol{\epsilon}_{c^d} \rfloor_\times\right). \tag{3.19}$$

If we express the orientation error using the angle-axis notation, $\boldsymbol{J}_\phi$ would correspond to the commonly used formulation presented in [3]

$$\boldsymbol{J}_\phi = \boldsymbol{I}_3 - \frac{\theta}{2}\lfloor \boldsymbol{\varphi} \rfloor_\times + \left(1 - \frac{sinc(\theta)}{sinc^2(\theta/2)}\right)\lfloor \boldsymbol{\varphi} \rfloor_\times^2. \tag{3.20}$$

Although in both (3.16) and (3.20), $\boldsymbol{J}_\phi$ becomes singular when $\theta = \frac{\pi}{2} + k\pi$, the interaction Jacobian (3.9) is full rank when $\theta \neq \frac{\pi}{2} + k\pi$ and we can reach global asymptotic stability (see Sect. 3.3.1) under the strong hypothesis that all the pose parameters are perfect.

3.5 Image-Based Visual Servo (IBVS)

In IBVS, the controller uses as input (i.e., the generic variable $\boldsymbol{s}$ in (3.1)) the 2D image coordinates of our set of target features detected in the image. IBVS computes the control error in 2D pose space and converts this error to camera velocities with the image Jacobian (3.2), defined in the following.

This 2D error computation for IBVS methods allows them to be remarkably more robust to errors in calibration and image noise than in PBVS schemes (i.e., less than 10% of error in camera intrinsic parameters [15]). In contrast to IBVS, PBVS methods require camera pose estimations in 3D, which in turn depend also on feature detections.

3.5.1 *Image Jacobian*

Let us consider $\boldsymbol{p} = \begin{bmatrix} x \ y \ z \end{bmatrix}^\top$ be a static 3D point, expressed in the camera reference frame c.

The projection of $\boldsymbol{p}$ onto the image plane, according to the pin-hole model, is defined as in (2.17) by

$$\pi = \begin{bmatrix} u \\ v \end{bmatrix} = \boldsymbol{P}_f \frac{\boldsymbol{p}}{z}, \tag{3.21}$$

with $\boldsymbol{P}_f$ the projection matrix described in (2.22) and z the distance, measured along the camera optical axis, from the center to $\boldsymbol{p}$.

In camera coordinates, the motion of the target can be described by a translational velocity $\boldsymbol{v}$ and an angular velocity $\boldsymbol{\omega}$. Then, the velocity of each feature in the camera frame is given by the expression

$$\dot{\boldsymbol{p}} = -\boldsymbol{v} - \boldsymbol{\omega} \times \boldsymbol{p} \tag{3.22}$$

which corresponds to

$$\begin{bmatrix} \dot{x} \\ \dot{y} \\ \dot{z} \end{bmatrix} = \begin{bmatrix} -v_x - \omega_y\, z + \omega_z\, y \\ -v_y - \omega_z\, x + \omega_x\, z \\ -v_z - \omega_x\, y + \omega_y\, x \end{bmatrix}. \tag{3.23}$$

Combining (3.23) with the time derivative of $\boldsymbol{\pi}$, (2.18), we obtain

$$\dot{u} = -\frac{v_x}{z} + \frac{u\, v_z}{z} + u\, \nu\, \omega_x - (1 + u^2)\omega_y + \nu\, \omega_z \tag{3.24a}$$

$$\dot{\nu} = -\frac{v_y}{z} + \frac{\nu\, v_z}{z} + (1 + \nu^2)\omega_x - u\, \nu\, \omega_y - \nu\, \omega_z. \tag{3.24b}$$

where, for a feature j, can be written as

$$\dot{\boldsymbol{s}}_j = \boldsymbol{J}_j \boldsymbol{\vartheta}_c \tag{3.25}$$

with $\dot{\boldsymbol{s}}_j = \left[\dot{u}_j\ \dot{\nu}_j\right]^\top$, the image velocities of the feature j, and $\boldsymbol{\vartheta}_c = \left[\boldsymbol{v}^\top\ \boldsymbol{\omega}^\top\right]^\top$, the camera velocities. $\boldsymbol{J}_j$ is the image Jacobian for the j-th feature, and takes the form

$$\boldsymbol{J}_j = \begin{bmatrix} -\frac{1}{z} & 0 & \frac{u}{z} & u\,\nu & -(1+u^2) & \nu \\ 0 & -\frac{1}{z} & \frac{\nu}{z} & (1+\nu^2) & -u\,\nu & -u \end{bmatrix}. \tag{3.26}$$

Stacking these together, we get the image Jacobian for all n features

$$\boldsymbol{J} = \begin{bmatrix} \boldsymbol{J}_1 \\ \vdots \\ \boldsymbol{J}_n \end{bmatrix}. \tag{3.27}$$

3.6 Uncalibrated Image-Based Visual Servo (UIBVS)

In all classical visual servo approaches (i.e., PBVS, IBVS and hybrid methods) the resulting interaction matrix or image Jacobian depends on a priori knowledge of the intrinsic camera parameters. Although image-based methods, and in extension some hybrid approaches, have shown more robustness in these parameters than pose-based approaches (as stated in the previous Section), they usually break down at error levels larger than 10% [15]. In the following, we present a method that indirectly estimates the focal length online which, as shown in the experiments section, allows to withstand much larger calibration errors. To do so, first we show some background formulation.

3.6.1 Background

Drawing inspiration on the EPnP [12] and UPnP [20] algorithms, we can formulate the focal length in terms of the relation between the camera and target frames. To this end, we set a reference system attached to the target object, and define a set of four control points as a basis for this reference system. Then, one can express the 3D coordinates of each target feature as a weighted sum of the elements of this basis. Computing the pose of the object with respect to the camera resorts to computing the location of these control points with respect to the camera frame. A least squares solution for the control point coordinates albeit scale, is given by the null eigenvector of a linear system made up of all 2D to 3D perspective projection relations between the target points. Given the fact that distances between control points must be preserved, these distance constraints can be used in a second least squares computation to solve for scale and focal length.

More explicitly, the perspective projection equations for each target feature, already described in (2.17), can now be expressed with

$$\sum_{j=1}^{4}\left(a_{ij}x_j + a_{ij}(u_0 - u_i)\frac{z_j}{\alpha}\right) = 0, \tag{3.28}$$

$$\sum_{j=1}^{4}\left(a_{ij}y_j + a_{ij}(v_0 - v_i)\frac{z_j}{\alpha}\right) = 0, \tag{3.29}$$

where $\boldsymbol{s}_i = [u_i, v_i]^\top$ are the image coordinates of the target feature i, and $\boldsymbol{c}_j = [x_j, y_j, z_j]^\top$ are the 3D coordinates of the j-th control point in the camera frame. The terms a_{ij} are the barycentric coordinates of the i-th target feature which are constant regardless of the location of the camera reference frame, and α is our unknown focal length.

These equations can be jointly expressed for all 2D-3D correspondences as a linear system

$$\boldsymbol{M}\boldsymbol{x} = \boldsymbol{0}, \tag{3.30}$$

where M is a $2n$x12 matrix made of the coefficients a_{ij}, the 2D points $\boldsymbol{s}_i$ and the principal point; and $\boldsymbol{x}$ is our vector of 12 unknowns containing both the 3D coordinates of the control points in the camera reference frame and the camera focal length, dividing the z terms:

$$\boldsymbol{x} = [x_1, y_1, z_1/\alpha, \ldots, x_4, y_4, z_4/\alpha]^T. \tag{3.31}$$

Its solution lies in the null space of $\boldsymbol{M}$, and can be computed as a scaled product of the null eigenvector of $\boldsymbol{M}^\top\boldsymbol{M}$ via Singular Value Decomposition

$$\boldsymbol{x} = \beta\boldsymbol{\mu}, \tag{3.32}$$

the scale β becoming a new unknown.

Notice how in the noise-free case, $\boldsymbol{M}^{\top}\boldsymbol{M}$ is only rank deficient by one, but when image noise is severe $\boldsymbol{M}^{\top}\boldsymbol{M}$ might loose rank, and a more accurate solution can be found as a linear combination of the basis of its null space. In this work, we consider only the least squares approximation. That is, only the eigenvector associated to the smallest eigenvalue.

To solve for β we add constraints that preserve the distance between control points of the form

$$||c_j - c_{j'}||^2 = d_{jj'}^2, \tag{3.33}$$

where $d_{jj'}$ is the known distance between control points c_j and $c_{j'}$ in the world coordinate system. Substituting $\boldsymbol{x}$ in the six distance constraints of (3.33), we obtain a system of the form

$$\boldsymbol{L}\boldsymbol{b} = \boldsymbol{d}, \tag{3.34}$$

where $\boldsymbol{b} = [\beta^2, \alpha^2\beta^2]^{\top}$, $\boldsymbol{L}$ is a 6×2 matrix built from the known elements of $\boldsymbol{\mu}$, and $\boldsymbol{d}$ is the 6-vector of squared distances between the control points. We solve this overdetermined linearized system using least squares and estimate the magnitudes of β and α by back substitution

$$\beta = \sqrt{b_1}, \tag{3.35a}$$

$$\alpha = \sqrt{\frac{|b_2|}{|b_1|}}. \tag{3.35b}$$

For a more exhaustive explanation of this method for pose and focal length estimation we refer the reader to the above-mentioned papers.

3.6.2 *Uncalibrated Image Jacobian*

We want to minimize the error shown in (3.1), where in this case we can select $\boldsymbol{s}$ to be the projection of the control points $\boldsymbol{c}$, and $\boldsymbol{s}^d$ their final desired position in the image plane, computed with our initial value for α.

Then, injecting (3.32) in the equations of the target motion (3.23), we obtain

$$\begin{bmatrix} \dot{x}_j \\ \dot{y}_j \\ \dot{z}_j \end{bmatrix} = \begin{bmatrix} -v_x - \omega_y \, \alpha \, \beta \mu_z + \omega_z \, \beta \mu_y \\ -v_y - \omega_z \, \beta \mu_x + \omega_x \, \alpha \, \beta \mu_z \\ -v_z - \omega_x \, \beta \mu_y + \omega_y \, \beta \mu_x \end{bmatrix}, \tag{3.36}$$

where μ_x, μ_y, and μ_z are the x, y, and z components of eigenvector $\boldsymbol{\mu}$ related to the control point $\boldsymbol{c}_j$ (3.32), and whose image projection is given by

$$\begin{bmatrix} u_j \\ v_j \end{bmatrix} = \begin{bmatrix} \alpha \frac{x_j}{z_j} + u_0 \\ \alpha \frac{y_j}{z_j} + v_0 \end{bmatrix}, \tag{3.37}$$

and its time derivative by

$$\begin{bmatrix} \dot{u}_j \\ \dot{v}_j \end{bmatrix} = \alpha \begin{bmatrix} \frac{\dot{x}_j}{z_j} - \frac{x_j \dot{z}_j}{z_j^2} \\ \frac{\dot{y}_j}{z_j} - \frac{y_j \dot{z}_j}{z_j^2} \end{bmatrix}. \tag{3.38}$$

Substituting (3.32) and (3.36) in (3.38) we obtain

$$\dot{u}_j = \frac{-v_x - \alpha\beta\mu_z\omega_y + \beta\mu_y\omega_z}{\beta\mu_z} - \frac{\mu_x(-v_z - \beta\mu_y\omega_x + \beta\mu_x\omega_y)}{\alpha\beta\mu_z^2} \tag{3.39}$$

$$\dot{v}_j = \frac{-v_y - \alpha\beta\mu_z\omega_x + \beta\mu_x\omega_z}{\beta\mu_z} - \frac{\mu_y(-v_z - \beta\mu_y\omega_x + \beta\mu_x\omega_y)}{\alpha\beta\mu_z^2}, \tag{3.40}$$

which can be rewritten as

$$\dot{\boldsymbol{s}}_j = \boldsymbol{J}_j \boldsymbol{\vartheta}_c, \tag{3.41}$$

with $\dot{\boldsymbol{s}}_j = [\dot{u}_j, \dot{v}_j]^\top$, the image velocities of control point j, and $\boldsymbol{\vartheta}_c = \left[\boldsymbol{v}^\top \; \boldsymbol{\omega}^\top\right]^\top$, the camera velocities. $\boldsymbol{J}_j$ is our seeked calibration-free image Jacobian for the j-th control point, and takes the form

$$\boldsymbol{J}_j = \begin{bmatrix} \frac{-1}{\beta\mu_z} & 0 & \frac{\mu_x}{\alpha\beta\mu_z^2} & \frac{\mu_x\mu_y}{\alpha\mu_z^2} & \frac{-\mu_x^2-\alpha^2\mu_z^2}{\alpha\mu_z^2} & \frac{\mu_y}{\mu_z} \\ 0 & \frac{-1}{\beta\mu_z} & \frac{\mu_y}{\alpha\beta\mu_z^2} & \frac{\mu_y^2+\alpha^2\mu_z^2}{\alpha\mu_z^2} & \frac{-\mu_x\mu_y}{\alpha\mu_z^2} & \frac{-\mu_x}{\mu_z} \end{bmatrix}. \tag{3.42}$$

Stacking these together, we get the image Jacobian for all control points

$$\boldsymbol{J} = \begin{bmatrix} \boldsymbol{J}_1 \\ \vdots \\ \boldsymbol{J}_4 \end{bmatrix}. \tag{3.43}$$

Notice how, the terms of $\boldsymbol{\mu}$ in our Jacobian are the coordinates, albeit scale of our control points, which in turn form a basis of the original features. They are by construction linearly independent and it can be shown that with such selection of control points, $\boldsymbol{J}$ has full rank 6 and thus, In Eq. 3.6 holds and thus asymptotic stability is guaranteed.

3.7 Validation

In this section, we show the advantages and limitations of the presented visual servo methods. Specifically we show the performance of the UIBVS approach by comparing it with the classical schemes (i.e., PBVS and IBVS) through simulation case studies, assuming a holonomic system with 6 DoFs. The following simulations were done in Matlab-Simulink using as template the Visual Servoing Toolbox,[2] and their

[2]http://vstoolbox.sourceforge.net/.

implementations are available online.[3] Some of these simulations are reported in Video 3, referenced in Appendix 5.A.

These visual-based servo methods were designed with a particular application in mind, that of maneuvering a UAM to a desired location for robotic manipulation tasks. As common UAM platforms are underactuated vehicles (i.e., 4 DoFs), we require extra DoFs to drive the camera with the visual servo velocities, e.g. attaching a serial arm of at least 2 DoFs. This kinematically augmented platforms are detailed in Chap. 4, and thus real experiments with the application of these visual servo methods are provided in Sect. 4.7.

Given a random set of target features, an initial camera position, and a desired final position with respect to the target, we want to compare the performance of the PBVS, IBVS and UIBVS algorithms with the same setup, except for the unknown camera calibration parameters in the uncalibrated case. For these comparisons we show simulation runs for 50 sec with time steps of 0.1 sec where the camera is assumed to be fully controllable (i.e., 6 DoF), and the controller used is a simple proportional controller with gain $\lambda = 0.125$. The value was chosen empirically to be able to compare the time both algorithms take to reduce both image and Cartesian errors.

Figure 3.3 shows the obtained 3D camera trajectories using all presented methods, as well as the corresponding control points. The camera is indicated with a yellow tetrahedron in the initial and final locations, and the camera trajectories are shown as a concatenation of camera frames, with their axes depicted in red, green, and blue colors. The original features and the control points are depicted with blue circles and red crosses, respectively.

Analizing their performances in 3D space, PBVS is the approach which performs better (i.e., with a trajectory close to a straight line) because its error is directly computed in 3D cartesian space. Although IBVS and UIBVS reach the target with a bit larger trajectories than PBVS, both methods are sufficient to accomplish the task. On the other hand, if we analize how the methods perform with the target tracking, it is expected to have better performances using image-based methods. This behavior is clearly seen in the corresponding image trajectories of the control points, shown in Fig. 3.4a, where the method with a shorter trajectory corresponds to the IBVS, whereas PBVS takes large image plane paths to align the control points. As all methods are based on an underlying features detector, using PBVS methods may imply a trajectory where these features lay outside the image plane, thus the target is lost and the task cannot be accomplished. For these reason, in the experiments section of Chap. 4 the PBVS scheme is only used inside a hybrid architecture where also an IBVS is applied.

The time evolution of errors in Cartesian coordinates and in the image plane of the control points is plotted in Fig. 3.5. Under equal noise-free simulation conditions, all methods have similar asymptotic convergence and compare adequately with respect to each other, reaching the goal at the desired pose with similar but not identical trajectories.

[3]https://gitlab.iri.upc.edu/asantamaria/VisualServo.

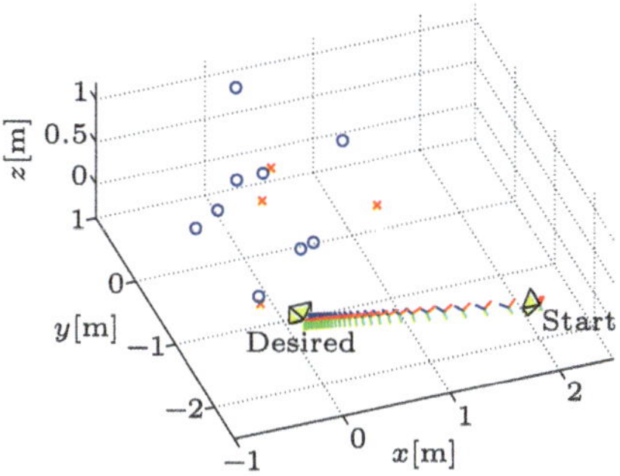

(a) Classical PBVS

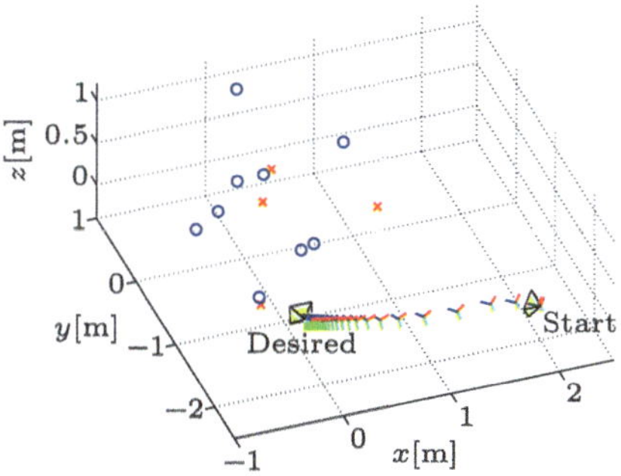

(b) Classical IBVS

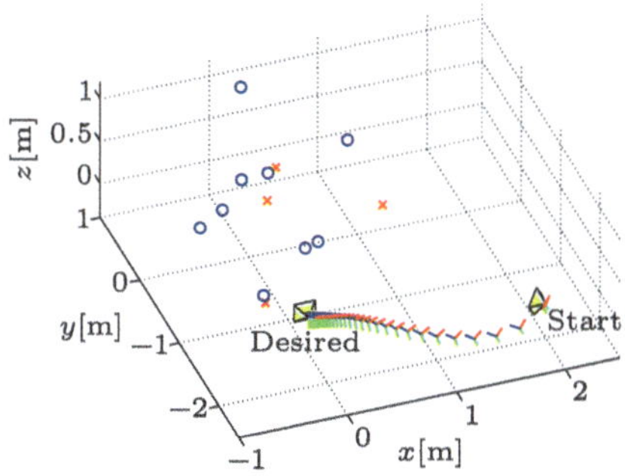

(c) UIBVS

Fig. 3.3 Camera trajectory comparison between **a** classical PBVS, **b** classical IBVS and **c** uncalibrated IBVS, under noise-free conditions

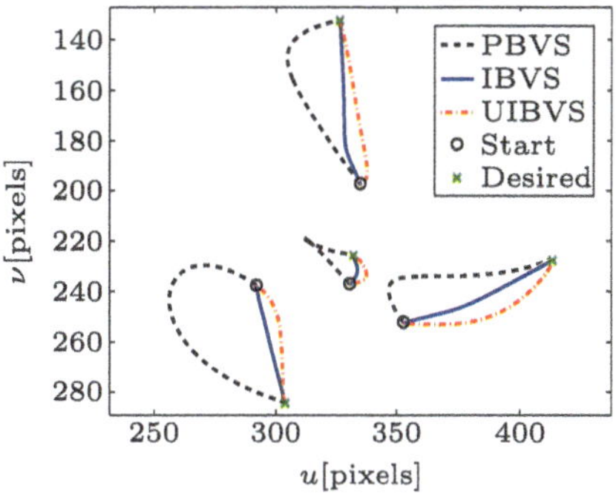

(a) Noise-free conditions

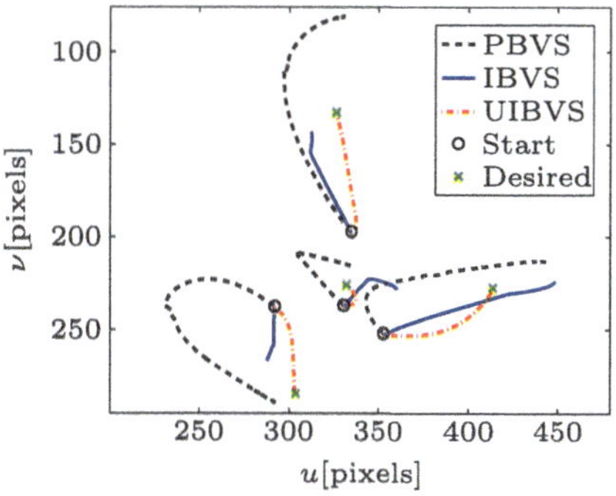

(b) Initialization error of 20% in camera focal length

Fig. 3.4 Comparison of control point trajectories in the image plane

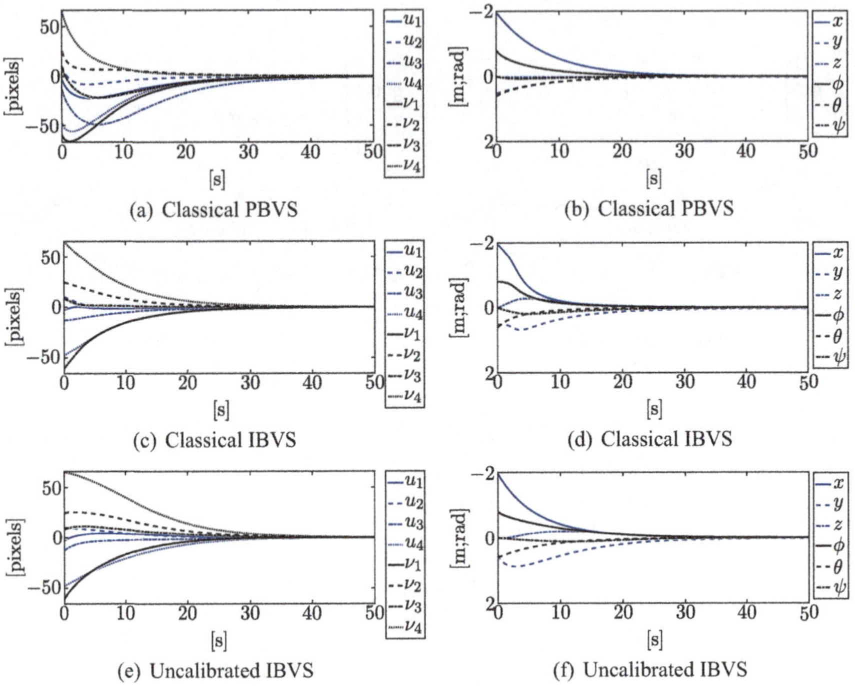

Fig. 3.5 Comparison of control point errors for noise-free conditions, **a–c** as reprojections in the image plane, and **d–f** as Cartesian coordinates in the camera reference frame

Fig. 3.6 Values of the Lyapunov candidate function $\mathscr{L}$ and its derivative $\dot{\mathscr{L}}$

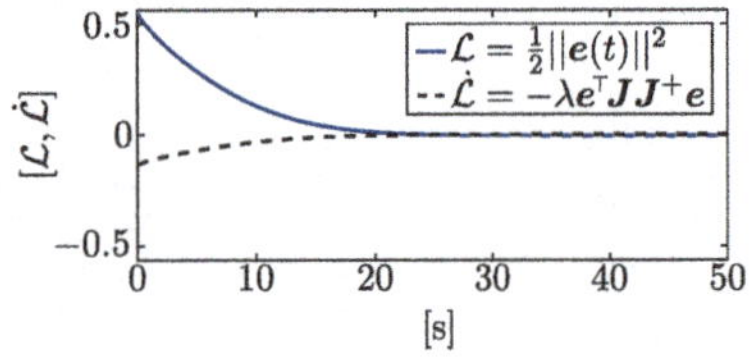

The presented UIBVS method is globally asymptotically stable when the sufficient condition in (3.6) holds. Our Jacobian is built from the coordinates, albeit scale, of our control points (3.43), which in turn form a basis of the original features. They are by construction linearly independent and it can be shown that with such selection of control points, $\boldsymbol{J}$ has full rank of dimension 6 and thus the inequality (3.6) holds. To show experimentally that our UIBVS scheme is globally asymptotically stable, we plot in Fig. 3.6 the value of the candidate function $\mathscr{L} = \frac{1}{2} \|\boldsymbol{e}(t)\|^2$, and its derivative $\dot{\mathscr{L}} = -\lambda \, \boldsymbol{e}^\top \boldsymbol{J} \boldsymbol{J}^+ \boldsymbol{e}$.

Now that our control scheme has been validated, we compare the method again versus the classical visual-based servo schemes, but now subject to noise, both in the image reprojections and in the internal camera parameters. IBVS and UIBVS methods turned out to be robust to noise levels of 1 to 3 pixels in the image coordinates but the interesting results were obtained when noise was added to the focal length, which can be caused by mechanical vibrations of the optics, bad initial calibration values, or unaccounted changes in zoom.

3.7.1 Noise in Focal Length

An unaccounted variation of focal length is assumed by the classical image-based servo approaches mainly as camera motion along the z axis. However, in the case of position-based methods, the focal length plays a role in the optimization process to obtain the camera pose from image feature detections and its impact is stronger. The major effect lies with the estimation of the camera orientation, because a small orientation error translates to large Cartesian positioning errors.

To recover from existing noise in the focal length, in all calibrated visual servo methods (i.e., PBVS, IBVS and HVS), the control law induces undesirable changes in the robot velocity commands. This is shown in Fig. 3.7, in which we plot the camera velocities for a servoing task with a focal length of 10 mm, and subject to white noise variations of 1 mm standard deviation. Plot Fig. 3.7a shows the computed camera velocities for the classical position-based method which is not able to drive

Fig. 3.7 Camera velocities during a servo task subject to white noise of 1 mm in the focal length

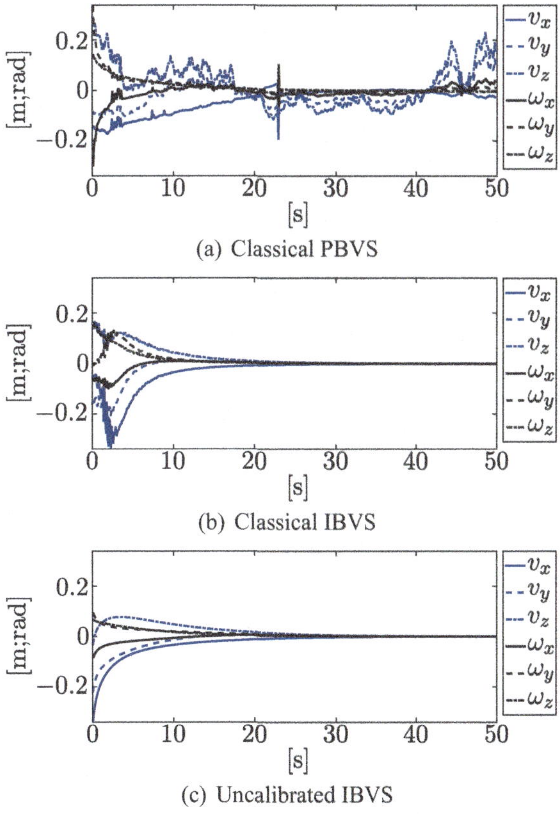

(a) Classical PBVS

(b) Classical IBVS

(c) Uncalibrated IBVS

the camera efficiently. The IBVS in plot Fig. 3.7b uses the focal-length dependent Jacobian. As stated in previous sections, image-based methods show more robustness to error in the camera intrinsic parameters than pose-based approaches [15]. Plot Fig. 3.7c corresponds to our proposed calibration-free scheme. Even when the servo task can be successfully completed in both image-based cases in approximately the same amount of time, the proposed method provides a much smoother tracking of the camera to such variations in focal length.

3.7.2 Wrong Initialization or Unaccounted Focal Length

The robustness of the UIBVS approach becomes more evident with large errors in the camera focal length, due to either a wrong calibration or for an unaccounted change. As shown previously, under equal noise-free simulation conditions all methods have comparable asymptotic convergence. But, for an unaccounted change in the focal length of 20% (e.g., a wrong initialization value), the classical PBVS and IBVS approaches are unable to reach the desired configuration, in contrast to the proposed approach in which the servoing task is completed without trouble. This behavior is show in Fig. 3.4b. Figure 3.8 shows the control point error trajectories, in the image plane (a–c), and in camera centered Cartesian coordinates (d–f).

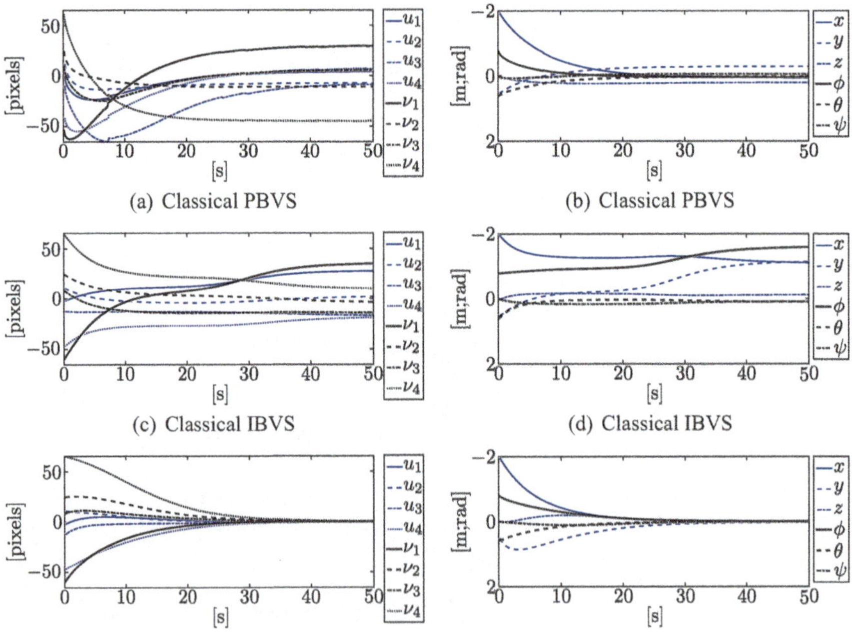

(a) Classical PBVS (b) Classical PBVS

(c) Classical IBVS (d) Classical IBVS

(e) Uncalibrated IBVS (f) Uncalibrated IBVS

Fig. 3.8 Comparison of control point errors for an unaccounted error of 20% in camera focal length, **a–c** as reprojections in the image plane, and (d-f) as Cartesian coordinates in the camera reference frame

3.8 Conclusions

In this chapter we described the principles of position- and image-based visual servo, together with a new method for uncalibrated cameras. As expected, image-based methods show more robustness to unaccounted changes in camera intrinsic parameters. However, if these errors are large (e.g., more than 10%) the classical image-based approach is not able to drive the camera to the target. Such errors are transfered to the control law, producing camera displacements along the optical axis. To overcome this situation, an uncalibrated visual servo method is presented, which shows robustness to such large errors in camera focal length because this distance is optimized on-line, thus allowing us to get rid of its dependency in the formulation of the image Jacobian.

In our uncalibrated image-based visual servo, target features are parametrized with their barycentric coordinates, and the basis of these coordinates is used to define a set of control points. A method is given to recover the coordinates of these control points and also of the camera focal length. With these, a new image Jacobian is derived which is guaranteed by construction to be of full rank. This guarantees asymptotic stability of the control law regardless of the target point selection, as long as planar configurations are avoided.

This research work has been partially published in [23] and [24]. All the technical details have been provided, facilitating the use of the proposed methods by other groups in the community. The techniques are here demonstrated in Matlab-Simulink and all our code is available for download.[4] A video of the method at work is also referenced in the same page and as Video 3 in Appendix 5.A.

We defined these visual servo tasks with a particular application in mind, that of driving a UAM for a realistic manipulation mission. Until now, in this chapter we assumed a holonomic system with 6DoF (i.e., the camera can move freely in space). In order to use these methods in a real UAM, we need to kinematically augment the multirotor platform with at least two DoFs due to its underactuation. In the next chapter we define such UAM, consisting on a multirotor platform with a serial arm attached below, and present experiments using the described visual servo approaches with real robot case studies.

References

1. Bourquardez, Odile, Robert Mahony, Nicolas Guenard, FranÇois Chaumette, Tarek Hamel, and Laurent Eck. 2009. Image-based visual servo control of the translation kinematics of a quadrotor aerial vehicle. *IEEE Transactions on Robotics* 25 (3): 743–749.
2. Chaumette, Francois. 1998. Potential problems of stability and convergence in image-based and position-based visual servoing. In *The confluence of vision and control*, number 237 in Lecture Notes in Control and Information Sciences, pp. 66–78. Springer.

[4]https://gitlab.iri.upc.edu/asantamaria/VisualServo.

3. Chaumette, François, and Seth Hutchinson. 2006. Visual servo control. I. Basic approaches. *IEEE Robotics & Automation Magazine* 13 (4): 82–90.
4. Chaumette, François, and Seth Hutchinson. 2007. Visual servo control. II. Advanced approaches. *IEEE Robotics & Automation Magazine* 14 (1): 109–118.
5. Corke, P.I., and S.A. Hutchinson. 2001. A new partitioned approach to image-based visual servo control. *IEEE Transactions on Robotics and Automation* 17 (4): 507–515.
6. Espiau, B., F. Chaumette, and P. Rives. 1992. A new approach to visual servoing in robotics. *IEEE Transactions on Robotics and Automation* 8 (3): 313–326.
7. Farahmand, Amir Massoud, Azad Shademan, and Martin Jägersand. 2007. Global visual-motor estimation for uncalibrated visual servoing. In *IEEE/RSJ international conference on intelligent robots and systems*, pp. 1969–1974, San Diego, USA.
8. Hashimoto, K., T. Kimoto, T. Ebine, and H. Kimura. 1991. Manipulator control with image-based visual servo. In *IEEE international conference on robotics and automation*, pp. 2267–2271, Sacramento, USA.
9. Hosoda, Koh, and Minoru Asada. 1994. Versatile visual servoing without knowledge of true jacobian. In *IEEE/RSJ international conference on intelligent robots and systems*, pp. 186–193, Munich, Germany.
10. Hutchinson, S., G.D. Hager, and P.I. Corke. 1996. A tutorial on visual servo control. *IEEE Transactions on Robotics and Automation* 12 (5): 651–670.
11. Janabi-Sharifi, Farrokh, Lingfeng Deng, and William J. Wilson. 2011. Comparison of basic visual servoing methods. *IEEE/ASME Transactions on Mechatronics* 16 (5): 967–983.
12. Lepetit, Vincent, Francesc Moreno-Noguer, and Pascal Fua. 2009. EPnP: An accurate O(n) solution to the PnP problem. *International Journal of Computer Vision* 81 (2): 155–166.
13. Lippiello, V., J. Cacace, A. Santamaria-Navarro, J. Andrade-Cetto, M. A. Trujillo, Y.R. Esteves, and A. Viguria. 2016. Hybrid visual servoing with hierarchical task composition for aerial manipulation. *IEEE Robotics and Automation Letters*, 1(1): 259–266. Presented in ICRA'16.
14. Lippiello, Vincenzo, Bruno Siciliano, and Luigi Villani. 2007. Position-based visual servoing in industrial multirobot cells using a hybrid camera configuration. *IEEE Transactions on Robotics* 23 (1): 73–86.
15. Malis, E., and P. Rives. 2003. Robustness of image-based visual servoing with respect to depth distribution errors. In *IEEE international conference on robotics and automation*, 1056–1061, vol. 1 China: Taipei.
16. Malis, Ezio, Francois Chaumette, and Sylvie Boudet. 1999. 2 1/2 D visual servoing. *IEEE Transactions on Robotics and Automation* 15 (2): 238–250.
17. Martinet, P., and J. Gallice. 1999. Position based visual servoing using a non-linear approach. In *IEEE/RSJ international conference on intelligent robots and systems*, vol. 1, pp. 531–536. Kyongju, Korea.
18. Mebarki, Rafik, Vincenzo Lippiello, and Bruno Siciliano. 2015. Nonlinear visual control of unmanned aerial vehicles in GPS-denied environments. *IEEE Transactions on Robotics* 31 (4): 1004–1017.
19. Mezouar, Youcef, and Francois Chaumette. 2003. Optimal camera trajectory with image-based control. *The International Journal of Robotics Research* 22 (10–11): 781–803.
20. Penate-Sanchez, Adrian, Juan Andrade-Cetto, and Francesc Moreno-Noguer. 2013. Exhaustive linearization for robust camera pose and focal length estimation. *IEEE Transactions on Pattern Analysis and Machine Intelligence* 35 (10): 2387–2400.
21. Piepmeier, Jenelle Armstrong, Gary V. McMurray, and Harvey Lipkin. 2004. Uncalibrated dynamic visual servoing. *IEEE Transactions on Robotics and Automation* 20 (1): 143–147.
22. Sanderson, A.C., and L.E. Weiss. 1980. Image-based visual servo control using relational graph error signals. *Proceedings of the IEEE* 68: 1074–1077.
23. Santamaria-Navarro, Angel, and Juan Andrade-Cetto. 2013. Uncalibrated image-based visual servoing. In *IEEE international conference on robotics and automation*, pp. 5247–5252, Karlsruhe, Germany.
24. Santamaria-Navarro, Angel, Patrick Grosch, Vincenzo Lippiello, Joan Solà, and Juan Andrade-Cetto. 2017. Uncalibrated visual servo for unmanned aerial manipulation. *Accepted for publication in the IEEE/ASME Transactions on Mechatronics*. To appear.

25. Shademan, Azad, Amir-Massoud Farahmand, and Martin Jägersand. 2010. Robust jacobian estimation for uncalibrated visual servoing. In *IEEE international conference on robotics and automation*, pp. 5564–5569, Anchorage, USA.
26. Thuilot, B., P. Martinet, L. Cordesses, and J. Gallice. 2002. Position based visual servoing: keeping the object in the field of vision. In *International conference on robotics and automation*, vol. 2, pp. 1624–1629, Washington DC, USA.
27. Viéville, Thierry, Cyril Zeller, and Luc Robert. 1996. Using collineations to compute motion and structure in an uncalibrated image sequence. *International Journal of Computer Vision* 20 (3): 213–242.
28. Wang, Yuquan, Johan Thunberg, and Xiaoming Hu. 2012. A transformation of the position based visual servoing problem into a convex optimization problem. In *IEEE conference on decision and control*, pp. 5673–5678, Maui, USA.
29. Westmore, D.B., and W.J. Wilson. 1991. Direct dynamic control of a robot using an end-point mounted camera and Kalman filter position estimation. In *IEEE international conference on robotics and automation*, pp. 2376–2384, Sacramento, USA.
30. Wilson, W.J., C.C. Williams Hulls, and G.S. Bell. 1996. Relative end-effector control using Cartesian position based visual servoing. *IEEE Transactions on Robotics and Automation* 12 (5): 684–696.

Chapter 4
Task Control

4.1 Introducion

Multirotors, and in particular quadrotors such as the ones used in this book, are underactuated platforms. That is, they can change their torque load and thrust/lift by altering the velocity of the propellers, with only four DoFs (e.g., one for the thrust and three torques). But, as shown in this Chapter, the attachment of a manipulator arm to the base of the robot can be seen as a strategy to alleviate underactuation allowing UAMs to perform complex tasks.

In this work, we attach a light-weight serial arm to a multirotor and use a camera, together with the visual servo methods presented in Chap. 3, to drive the UAM towards a desired target. The arm lets us exploit the redundancy in DoFs of the overall system not only to achieve the desired visual servo task, but to do so whilst attaining also other tasks during the mission, e.g. avoiding obstacles or self collisions, compensating for changes in weight distribution during arm operation, or driving the arm to a desired configuration with high manipulability, thus improving overall flight behavior.

Redundancy in DoFs is exploited by combining the tasks hierarchically so as to tackle additional objectives expressed as constraints. In this Chapter we define two hierarchical methods. A first approach, presented in [29], addresses a full least squares secondary task solution by suitably assigning an order of priority to the given tasks and then satisfying the lower-priority task only in the null space of the higher-priority task. A second method, presented in [51, 52], combines tasks hierarchically in a less restrictive manner than [29], minimizing secondary task reconstruction only for those components not in conflict with the primary task. This strategy is known to achieve possibly less accurate secondary task reconstruction but with the advantage of decoupling algorithmic singularities between tasks [1]. Although hierarchical task composition techniques are well known for redundant manipulators, its use on aerial manipulation is novel. In both cases, we address the problem of coupling two different systems, an aerial platform with a serial arm, where the underactuation of the flying vehicle has critical effects on mission achievement. Therefore, we show how the non-controllable DoFs must be considered in the task designs.

A. Santamaria-Navarro et al., *Visual Guidance of Unmanned Aerial Manipulators*, Springer Tracts in Advanced Robotics 125,
https://doi.org/10.1007/978-3-319-96580-2_4

An optimization based approach can also be used to exploit redundancy. We also present in this Chapter a quadratic programming method to compute joint trajectories for UAMs [49]. In order to alleviate the optimization procedure, the relative importance of tasks is set using a weighting strategy, leading to an on-board and real-time method. In this case, the definition of the problem is in the acceleration domain, which allows us not only to integrate and perform a large set of tasks, but also to obtain smooth motion of the joints.

In this Chapter we also provide the definitions of the previously mentioned tasks, specially designed for UAMs. Specifically, we define a safety task intended for obstacle and self-collision avoidance. Then, to minimize undesired effects of the arm onto the platform, we describe a task to vertically align the arm center of gravity with that of the platform, and another one to limit the forces exerted by the robotic arm on the quadrotor horizontal plane. To improve arm motion, we define two well known tasks consisting on maximizing the arm manipulability and to reach desired arm joint positions. We also present a task to limit the quadrotor accelerations and, to assure the convexity of the problem in the case of using the optimization technique, we describe a velocity minimization cost function.

We validate the use of the proposed hierarchical control laws and the optimization technique in simulation case studies and in extensive real experiments.

The remainder of this Chapter is structured as follows. The next section overviews the state of the art in aerial manipulation with special attention to UAM control. Section 4.2 describes the kinematics of two UAM configurations. Then we describe the task control in Sect. 4.3, with the proposed hierarchical formulation defined in Sect. 4.4, and the quadratic programming solution in Sect. 4.5. Section 4.6 defines the tasks designed for UAMs. Simulations and experimental results are presented in Sect. 4.7. Finally, conclusions and a summary of contributions are given in Sect. 4.8.

The modeling and control of a UAV able of interacting with the environment to accomplish simple robotic-manipulation tasks have been proposed in [31], which is based on a force-position control law designed through a feedback linearizing technique. In [17] an analogous issue has been investigated by further considering a safe take-off from forbidding terrains for vertical take-off and landing (VToL) UAV. With the improvement of the batteries and the miniaturization of motors and servos, new high-performance UAV prototypes endowed with a robot arm have been designed (i.e., a UAM).

The ability for small UAVs to manipulate or carry objects could greatly expand the types of missions achievable by such unmanned systems. High performance arms with end effectors typically weigh more than 10 kg, which cannot be supported by most commercially available small-sized UAVs. In contrast, aerial manipulation tasks executed with a UAM endowed with a simple light-weight robot arm with 2 DoFs is presented in [21], where an adaptive sliding-mode controller has been adopted, leading to a 6 DoF positioning system (without overactuation). Recent developments however suggest a trend change with UAV payload capabilities increasing and arm weights getting smaller [25, 41, 43].

Flying with a suspended load is a challenging task since the vehicle is characterized by unstable dynamics in which the presence of the object causes nontrivial coupling effects and the load significantly changes the flight characteristics. Given that the stability of the vehicle-load system must be preserved, it is essential for the flying robot to have the ability to minimize the effects of the arm in the flying system during the assigned maneuvers [45]. Kondak et al. [23] and Huber et al. [19] show the effect of using independent controllers for both the arm and the aerial platform. They take advantage of an industrial manipulator attached to a main-tail-rotor helicopter (120 kg of payload), and to improve the kinematic coupling, they suggest to use the platform yaw axis as an extra arm DoF in order to achieve simple manipulation tasks.

Among the undesired dynamic effects, there is the change of the center of mass during flight, that can be solved designing a low-level attitude controller such as a Cartesian impedance controller. In [27, 30] the dynamic model of a UAM and a Cartesian impedance control have been designed providing a desired relationship between external wrench and the system motion. However, redundancy is exploited in a rigid way. A different approach is [44], where an adaptive controller is set based on output feedback linearization to compensate the unknown displacement of the center of mass during aggressive maneuvers. However, only the case of a quadrotor is studied.

In contrast, [4] address this problem designing a light-weight arm with a differential joint at the base of the robotic arm and relocating the rest of the motors also in the arm base to reduce the misalignment of the arm CoG. Moreover, a desired end effector pose might require a non-horizontal robot configuration that the low level controller would try to compensate, changing in turn the arm end effector position. In this way, [42] design a controller exploiting the manipulator and quadrotor models. However flight stability is preserved by restricting the arm movements to those not jeopardizing UAM integrity. The dynamic stability of a UAM under the influence of grasped objects has been addressed in [43].

A control algorithm which is able to exploit all the DoFs of a UAM is proposed in [14], where the execution of tasks with a physical interaction with the environment has been achieved. The employed UAM is completely actuated only along one direction. A related control solution considering valve turning with a dual-arm UAM was proposed in [24]. In a different approach, [6, 38] consider the coordination of UAVs transporting a payload via cables, where robot motions are generated ensuring static equilibrium of the load. In all these works, vision is not employed for task execution and there is no redundancy. The redundancy of the system in the form of extra DoFs can be exploited to develop lower priority stabilizing tasks after the primary task, which can be simultaneously performed within a hierarchical framework, by optimizing some given quality indices, e.g. manipulability, joint limits, etc., [3, 11].

The use of vision for the execution of aerial robotic tasks is a widely adopted solution to cope with unknown environments. In [33–35] new image-based control laws are presented to automatically position UAM parts on target structures, where the system redundancy and underactuation of the vehicle base are explicitly taken into account. The camera is attached on the aerial platform and the positions of both the arm end effector and the target are projected onto the image plane in order to perform an image-based error decrease. When projecting the end effector, both end effector and vehicle velocities are required to be known, which for the second case creates a dependency on the robot odometry estimator that rarely achieves the required precision for aerial manipulation in a real scenario (e.g., without motion capture systems). Moreover, in both [34, 35] the proposed control schemes are only validated in simulation.

In [22] presents a vision-based method to guide a multirotor vehicle with a three DoFs arm attached. A traditional image-based approach is used to retrieve desired camera velocities, and to drive the whole system an adaptive controller is designed considering both kinematic and dynamic models. Although the complete robot has an extra DoF (i.e., 4 DoFs for the platform and 3 DoFs for the robot arm), this overactuation is not explicitly exploited. To cope with underactuation of the aerial platform, roll and pitch motion compensation is moved to the image processing part, requiring projective transformations and entailing some constraints. The errors computing arm kinematics are to be coupled with the image-based control law (this can be an important issue in case of realistic operations with aerial manipulators where the robustness of the joints are restricted by their payload). Moreover, the scale (i.e., the distance from the camera to the observed object) is strictly required and cannot be directly measured, thus some assumptions must be made or other sensors are to be used.

We presented in [52] a task-oriented control law for aerial surveillance, where a camera is attached to the end effector of the robot arm to perform visual servoing towards a desired target. However in that paper we employed redundancy in a rigid way and the interaction between dependent tasks was not considered.

In [29] we presented a hybrid position- and image-based servo scheme with hierarchical task composition for unmanned aerial manipulation that addresses full least squares secondary task solution. The corresponding control law is one of the contributions of this Chapter and is described in Sect. 4.4. The presence of redundancy in a UAM system allows combining a number of subtasks with a hierarchical-task formulation. Different subtasks can be designed both in the Cartesian space (e.g., obstacle avoidance or manipulation tasks), in the image space of the camera (e.g., IBVS or FoV constraints), as well as in the arm joint space (e.g., CoG balancing, joint-limits avoidance, manipulability, etc.). Moreover, the underactuation of the aerial vehicle base was systematically taken into account within a new recursive formulation. Although a close approach is [7], in [29] we derived a new advanced formulation with the capability to guarantee decoupling of independent tasks (not only orthogonal as in the previous work), the stability analysis of the new proposed control law is discussed together with the derivation of all the task Jacobian matrices (in [7] the Jacobian matrices of the uncontrollable variables are missing). Similarly, another contribution of this Chapter is our work presented in [51], described in Sect. 4.4, which formulates a similar control law but this time only requiring independence of non-controllable DoFs in the tasks to guarantee stability.

Even when the state of the art in control algorithms for UAMs is extensive, solutions for trajectory generation using optimal control in real-time are rare. These methods usually require powerful computational units due to their iterative nature. Hehn and D'Andrea. [18] and Mellinger and Kumar [36] describe methods for 3D optimal trajectory generation and control, however their works are focused only on

UAVs, thus considering few DoFs. In [53], trajectory generation is optimized for a vehicle with a cable-suspended load computing nominal trajectories with various constraints regarding the load swing. The application of such optimal control for UAMs can be seen in [15], where a nonlinear model predictive control scheme is proposed to achieve pick-and-place operations. Another example is [16] where a linear model predictive control is described using a direct multiple shooting method. However, results for UAMs are only shown in a simulated environment.

Drawing inspiration from other robotic fields (e.g., [12, 26]), in our work described in Sect. 4.5 (also presented in [49]) we take advantage of a quadratic programming technique to solve several UAM tasks in real time, on board and subject to constraints. Specifically we use the on-line active set strategy ([13, 47]) which analyses the constraints that are active at the current evaluation point, and give us a subset of inequalities to watch while searching for the solution, which reduces the complexity of the search and thus the computation time. A similar approach is [56], where the common idea consists in dividing the problem in two parts, firstly accounting for the trajectory generation and then implementing a reactive controller guaranteeing bounds on velocities and accelerations and enforcing a hierarchical structure of constraints. In contrast to [56], in which the experimental case study is based on a 7 DoF serial arm, we present a similar technique with specific tasks, constraints and bounds designed for UAMs. In this case, trajectory generation and redundancy exploitation are integrated in the same framework. To our knowledge, this has been the first work applied to such robots with all computations done in real time, and on board a limited computational unit.

4.2 Robot Kinematics

Traditionally, when performing visual servoing techniques with grounded serial arms, the hardware configurations of the arm end effector (hand) and the camera have been classified in two fundamental groups [10]:

- *Eye-in-hand*, or end-point closed-loop control, where the camera is attached to the moving hand and observing the relative position of the target.
- *Eye-to-hand*, or end-point open-loop control, where the camera is fixed in the world and observes both the target and the motion of the hand.

With the appearance of UAMs a new classification called *onboard-eye-to-hand* arose [33] where the camera is mounted on board the aerial platform (onboard-eye) while observing the robot manipulator (eye-to-hand). In this Chapter both *onboard-eye-to-hand* and *eye-in-hand* taxonomies are used to drive the UAM end effector using visual information towards a target in the scene. These hardware configurations are schematized in Fig. 4.1 and their coordinate frames and kinematic particularities are detailed in the following.

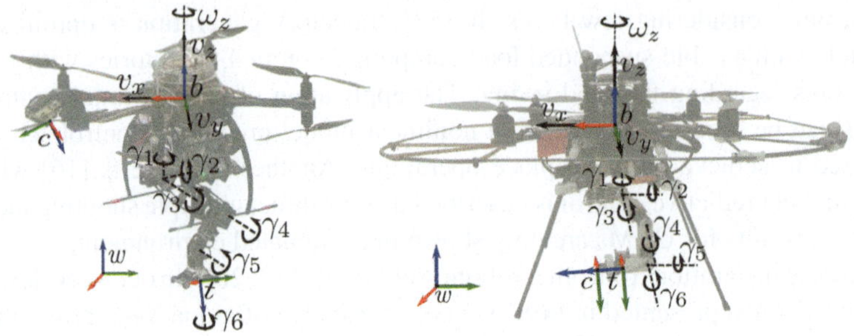

(a) Bonebreaker robot (1 : 20 scale). (b) Kinton robot (1 : 10 scale).

Fig. 4.1 The UAMs from the ARCAS project used in the experiments are composed of a 4 DoFs multirotor, commanded at a high-level by three translational and one angular velocities (v_x,v_y,v_z and ω_z), and a 6 DoFs robotic arm with joints γ_κ, $\kappa = 1 \ldots 6$; and world, camera, tool and body reference frames indicated by the letters w, c, t and b, respectively. Frame **a** corresponds to the *onboard-eye-in-hand* configuration while frame **b** is the *eye-in-hand* setting

4.2.1 *State Vector and Coordinate Frames*

Without loss of generality, we consider an inertial world frame w to be located at the target. Our goal is to operate with the tool attached at the arm end effector and defined by the frame t. With this, we define the camera frame c, whose pose with respect to the world frame w, expressed as a homogeneous transform ${}^w\boldsymbol{T}_c$, can be computed integrating the camera velocities obtained from one of the visual servo approaches presented in Chap. 3.

Let us define a complete system state vector $\boldsymbol{x}$ with

$$\boldsymbol{x} = \left[\boldsymbol{p}^\top \ \boldsymbol{\phi}^\top \ \boldsymbol{\gamma}^\top\right]^\top = \left[p_x \ p_y \ p_z \ \phi \ \theta \ \psi \ \gamma_1 \ldots \gamma_m\right]^\top , \tag{4.1}$$

where $\boldsymbol{p} \in \mathbb{R}^3$ and $\boldsymbol{\phi} \in \mathbb{R}^3$ are the position and orientation of the body frame b expressed in the inertial frame w, and $\boldsymbol{\gamma} \in \mathbb{R}^m$ are the arm joint angles, with m the number of arm DoFs. Similarly, we can define a complete twist vector $\boldsymbol{\xi}$ being

$$\boldsymbol{\xi} = \left[\boldsymbol{\vartheta}^\top \ \dot{\boldsymbol{\gamma}}^\top\right]^\top = \left[v_x \ v_y \ v_z \ \omega_x \ \omega_y \ \omega_z \ \dot{\gamma}_1 \ldots \dot{\gamma}_m\right]^\top , \tag{4.2}$$

where $\boldsymbol{\vartheta} = \left[\boldsymbol{v}^\top \ \boldsymbol{\omega}^\top\right]^\top$ is the platform twist, with $\boldsymbol{v} \in \mathbb{R}^3$ and $\boldsymbol{\omega} \in \mathbb{R}^3$ the translational and angular velocities of the body expressed in its own body frame b, and $\dot{\boldsymbol{\gamma}} \in \mathbb{R}^m$ are the arm joint angular velocities.

Notice how the twist vector ($\boldsymbol{\xi}$) is not directly the time derivative of the system state vector ($\dot{\boldsymbol{x}}$). This appreciation is important depending on which control references are considered when globally controlling the multirotor. Individually, each rotation rate (i.e., $\dot{\phi}, \dot{\theta}$ and $\dot{\psi}$) give us the respective angular velocity (e.g., $\omega_z = \dot{\psi}$), however they do not form an orthogonal set (i.e., a set of Euler angles do not constitute a Lie group in $\mathscr{SO}(3)$). When using Euler angles, the conversion of angular rates and angular velocities between two frames is defined by the so-called Wronskian

$$\boldsymbol{\omega} = \boldsymbol{\Gamma}\dot{\boldsymbol{\phi}} = \begin{bmatrix} 1 & 0 & -s_\theta \\ 0 & c_\phi & s_\phi c_\theta \\ 0 & -s_\phi & c_\phi c_\theta \end{bmatrix} \dot{\boldsymbol{\phi}} . \tag{4.3}$$

with the notation $s_x = \sin(x)$, $c_x = \cos(x)$, $t_x = \tan(x)$. Although the inverse matrix $\boldsymbol{\Gamma}^{-1}$ becomes singular with $\theta = \frac{\pi}{2} + k\pi$, we consider that UAMs are not meant for acrobatic maneuvers and these values will not be reached by the multirotor. In this Chapter, we provide high-level task control laws using local information from a camera to control $\boldsymbol{\xi}$, thus the use of this Wronskian is considered included in the low-level attitude control and does not appear in the following kinematics formulation.

We define the control velocities of the multirotor in the robot body frame b. A multirotor is at the high level of control an underactuated vehicle with only 4 DoFs, namely the linear velocities plus the yaw angular velocity (v_x, v_y, v_z and ω_z) acting on the body frame. Then, we can define a vector $\boldsymbol{\rho}$ containing only the controllable UAM DoFs with

$$\boldsymbol{\rho} = \begin{bmatrix} p_x & p_y & p_z & \psi & \gamma_1 & \dots & \gamma_m \end{bmatrix}^\top , \tag{4.4}$$

and its derivative

$$\dot{\boldsymbol{\rho}} = \begin{bmatrix} v_x & v_y & v_z & \omega_z & \dot{\gamma}_1 & \dots & \dot{\gamma}_m \end{bmatrix}^\top . \tag{4.5}$$

The pose of the multirotor body with respect to the target is determined by the homogenous transform ${}^w\boldsymbol{T}_b = {}^w\boldsymbol{T}_c {}^b\boldsymbol{T}_c^{-1}$, with ${}^w\boldsymbol{T}_c$ the camera pose expressed in the world reference frame and ${}^b\boldsymbol{T}_c$ the transform between body and camera frames that will depend on which configuration do we use (i.e., *onboard-eye-in-hand* or *eye-in-hand*), as explained in the following. To make it clear to the reader, left superscripts indicate the reference frame.

With an *onboard-eye-in-hand* setting, the camera is rigidly attached to the platform and decoupled from the arm motion. In this case, the arm configuration does not play a role when translating camera velocities from the visual servo to platform commands. Instead, to operate with the tool installed in the arm end effector the arm kinematics must be considered in the specific end effector task (e.g., gripper positioning with respect to the target). Being ${}^b\boldsymbol{T}_t(\boldsymbol{\gamma})$ the arm kinematics, for an *onboard-eye-in-hand* configuration we have ${}^w\boldsymbol{T}_t = {}^w\boldsymbol{T}_c {}^b\boldsymbol{T}_c^{-1} {}^b\boldsymbol{T}_t$. Notice how the arm base frame is considered, without loss of generality, coincident with the multirotor body frame.

Instead, with an *eye-in-hand* setting where the camera is rigidly attached at the arm end effector (${}^t\boldsymbol{T}_c$), to translate the visual servo velocities to the multirotor body frame we have to consider the relation ${}^b\boldsymbol{T}_c = {}^b\boldsymbol{T}_t {}^t\boldsymbol{T}_c$. In this configuration, the movement of the tool is directly expressed with ${}^w\boldsymbol{T}_t = {}^w\boldsymbol{T}_c ({}^t\boldsymbol{T}_c)^{-1}$.

Now, we can define a robot Jacobian that relates the local translational and angular velocities of the platform and those of the m arm joints ($\boldsymbol{\xi}$), to the desired camera velocities, namely ${}^{c}\boldsymbol{\vartheta}_{C}$, computed from the visual servo, such as

$$ {}^{c}\boldsymbol{\vartheta}_{C} = \boldsymbol{J}_{R}\,\boldsymbol{\xi}. \tag{4.6} $$

with $\boldsymbol{J}_{R}$ the Jacobian matrix of the whole robot. To make it clear to the reader, capital letters in subscripts indicate a referenced element with R, P, A, C and T resulting the whole UAM robot, multirotor platform, robot arm, camera and tool respectively. The kinematic relationships for both UAM configurations are explained in the following sections.

4.2.2 *Onboard-Eye-to-Hand Kinematics*

Let us consider the multirotor-arm system equipped with a camera mounted on the multirotor platform in Fig. 4.1a. With this robot setting, the arm motion does not influence the camera motion, thus the arm kinematics are not involved in the formulation of $\boldsymbol{J}_{R}$ when using local information (i.e., camera detections). For this reason, and considering that a main task for UAMs is to manipulate with its end effector, we present here not only the formulation of the Jacobian $\boldsymbol{J}_{R}$ but also the kinematic relations to translate global end effector velocities to those of the robot joints (4.5).

When the camera is subject to motion, its velocities defined in camera frame c can be directly expressed as a function of the platform translational and angular velocities in body coordinates with

$$ {}^{c}\boldsymbol{\vartheta}_{C} = \begin{bmatrix} {}^{c}\boldsymbol{R}_b & \boldsymbol{0}_3 \\ \boldsymbol{0}_3 & {}^{c}\boldsymbol{R}_b \end{bmatrix} \begin{bmatrix} \boldsymbol{v} + \boldsymbol{\omega} \times {}^{b}\boldsymbol{p}_{c} \\ \boldsymbol{\omega} \end{bmatrix} = {}^{c}\overline{\boldsymbol{R}}_b \begin{bmatrix} \boldsymbol{I}_3 & -\lfloor {}^{b}\boldsymbol{p}_c \rfloor_{\times} \\ \boldsymbol{0} & \boldsymbol{I}_3 \end{bmatrix} \boldsymbol{\vartheta}, \tag{4.7} $$

where ${}^{b}\boldsymbol{p}_{c}$ is the position of the camera, and ${}^{c}\boldsymbol{R}_b$ encodes the rotation between body and camera frames. Notice how in this formulation $\boldsymbol{\vartheta} \in \mathbb{R}^6$, including the non controllable DoFs of the platform. Then, the robot Jacobian for this configuration can be extracted by direct inspection of (4.7), being

$$ \boldsymbol{J}_{R} = \begin{bmatrix} {}^{c}\overline{\boldsymbol{R}}_b \begin{bmatrix} \boldsymbol{I}_3 & -\lfloor {}^{b}\boldsymbol{p}_c \rfloor_{\times} \\ \boldsymbol{0} & \boldsymbol{I}_3 \end{bmatrix} & \boldsymbol{0}_{6\times m} \end{bmatrix}. \tag{4.8} $$

Using this setting, the camera motion is only produced by platform movements—see the $\boldsymbol{0}_{6\times m}$ element in (4.8). Hence, considering the underactuaction of multirotors, we cannot completely satisfy 6 DoFs camera velocity references from a visual servo approach.

A possible solution is the use of an hybrid visual servo architecture as shown in Sects. 4.6.2 and 4.6.3, combining a PBVS for end effector trajectory tracking and IBVS to keep the target in the FoV. Hence, the arm kinematics, in the form of the well known arm Jacobian $\boldsymbol{J}_{A}$, will only appear in the definition of those tasks that pretend to specifically move the UAM end effector. As an example, we describe in the following the kinematics involved in a global end effector tracking task.

Let us consider an end effector velocity vector expressed in the world frame w, namely ${}^{w}\boldsymbol{\vartheta}_{T}$. We can define a robot Jacobian that relates the local translational and angular velocities of the platform and those of the m arm joints, $\boldsymbol{\xi}$, such as ${}^{w}\boldsymbol{\vartheta}_{T} = \boldsymbol{J}_{R}\boldsymbol{\xi}$. To do so, we can decompose ${}^{w}\boldsymbol{\vartheta}_{T}$ in the contributions of the platform and those of the arm with

$$ {}^{w}\boldsymbol{\vartheta}_{T} = {}^{w}\boldsymbol{\vartheta}_{T_P} + {}^{w}\boldsymbol{\vartheta}_{T_A} \,. \tag{4.9} $$

Following a similar procedure as in (4.7), it is easy to obtain

$$ {}^{w}\boldsymbol{\vartheta}_{T_P} = {}^{w}\overline{\boldsymbol{R}}_b \begin{bmatrix} \boldsymbol{I}_3 & -\lfloor {}^{b}\boldsymbol{p}_w \rfloor_{\times} \\ \boldsymbol{0} & \boldsymbol{I}_3 \end{bmatrix} \boldsymbol{\vartheta} \,. \tag{4.10} $$

The end effector velocities produced by the arm movement are described by

$$ {}^{w}\boldsymbol{\vartheta}_{T_A} = {}^{w}\overline{\boldsymbol{R}}_b \, \boldsymbol{J}_{A} \, \dot{\boldsymbol{\gamma}} \,, \tag{4.11} $$

with $\boldsymbol{J}_{A}$ the arm Jacobian. In summary, a global end effector velocity can be transformed to platform and arm joint velocities with

$$ \boldsymbol{J}_{R} = \begin{bmatrix} {}^{w}\overline{\boldsymbol{R}}_b \begin{bmatrix} \boldsymbol{I}_3 & -\lfloor {}^{b}\boldsymbol{p}_w \rfloor_{\times} \\ \boldsymbol{0} & \boldsymbol{I}_3 \end{bmatrix} & {}^{w}\overline{\boldsymbol{R}}_b \, \boldsymbol{J}_{A} \end{bmatrix} . \tag{4.12} $$

4.2.3 *Eye-in-Hand Kinematics*

With a multirotor-arm system where the camera is mounted at the end effector's arm as shown in Fig. 4.1b, we can follow a similar procedure as in (4.12). The camera velocities can be expressed as ${}^{c}\boldsymbol{\vartheta}_{C} = \boldsymbol{J}_{R}\,\boldsymbol{\xi}$, with

$$ \boldsymbol{J}_{R} = \begin{bmatrix} {}^{c}\overline{\boldsymbol{R}}_b \begin{bmatrix} \boldsymbol{I}_3 & -\lfloor {}^{b}\boldsymbol{p}_c \rfloor_{\times} \\ \boldsymbol{0}_3 & \boldsymbol{I}_3 \end{bmatrix} & {}^{c}\overline{\boldsymbol{R}}_b \, \boldsymbol{J}_{A} \end{bmatrix} . \tag{4.13} $$

4.3 Task Control

Even though the multirotor itself is underactuated (4 DoFs), by attaching a robotic arm with more than 2 DoFs we can attain over-actuation ($n = 4 + m$). In our case, $m = 6$. Exploiting this redundancy, we can achieve additional tasks acting on the null space of the robot Jacobian [40], while preserving the primary task. These tasks can be used to reconfigure the robot structure without changing the position and orientation of the arm end effector. This is usually referred to as *internal motion* of the arm.

One possible way to specify a secondary task is to choose its velocity vector as the gradient of a scalar objective function to optimize [11, 39]. Multiple secondary tasks can be arranged in hierarchy and, to avoid conservative stability conditions [2], the *augmented inverse-based projections* method is here considered [3]. In this method, lower priority tasks are not only projected onto the null space of the task up in the hierarchy, but onto the null space of an augmented Jacobian with all higher priority tasks.

In a general sense, we can define any such primary task as a configuration dependent task $\boldsymbol{\sigma}_0 = \boldsymbol{f}_0(\boldsymbol{x})$, with $\boldsymbol{x}$ the complete system state vector from (4.1). Differentiating it with respect to $\boldsymbol{x}$, and separating the uncontrollable state variables (i.e., roll and pitch in the case of a multirotor platform) we have

$$\dot{\boldsymbol{\sigma}}_0 = \frac{\partial \boldsymbol{f}_0(\boldsymbol{x})}{\partial \boldsymbol{x}} \dot{\boldsymbol{x}} = \boldsymbol{J}_0 \dot{\boldsymbol{\rho}}_0 + \overline{\boldsymbol{J}}_0 \boldsymbol{\varpi} , \tag{4.14}$$

where $\boldsymbol{\varpi} = \left[\omega_x, \omega_y\right]^\top$, $\overline{\boldsymbol{J}}_0$ is the Jacobian formed by the columns of the robot Jacobian corresponding to ω_x and ω_y, and $\boldsymbol{J}_0$ is the Jacobian formed by all other columns of $\boldsymbol{J}_R$, corresponding to the actuated variables $\dot{\boldsymbol{\rho}}$ from (4.5).

By inverting (4.14) and considering a regulation problem of $\boldsymbol{\sigma}_0$ to the desired value $\boldsymbol{\sigma}_0^d$, hence by defining $\widetilde{\boldsymbol{\sigma}}_0 = \boldsymbol{\sigma}_0^d - \boldsymbol{\sigma}_0$ as the main task error, the velocity command for the main task becomes

$$\dot{\boldsymbol{\rho}}_0 = \boldsymbol{J}_0^+ (\boldsymbol{\Lambda}_0 \widetilde{\boldsymbol{\sigma}}_0 - \overline{\boldsymbol{J}}_0 \boldsymbol{\varpi}) = \boldsymbol{J}_0^+ \boldsymbol{\varepsilon}_0 , \tag{4.15}$$

where $\boldsymbol{\Lambda}_0$ is a square positive-definite gain matrix, and $\boldsymbol{J}_0^+ = (\boldsymbol{J}_0^\top \boldsymbol{J}_0)^{-1} \boldsymbol{J}_0^\top$ is the left Moore-Penrose pseudoinverse of $\boldsymbol{J}_0$, which has been assumed to be full-rank.

By substituting (4.15) into (4.14), gives

$$\dot{\boldsymbol{\sigma}}_0 = \boldsymbol{\Lambda}_0 \widetilde{\boldsymbol{\sigma}}_0 , \tag{4.16}$$

which for a defined main task error $\widetilde{\boldsymbol{\sigma}}_0 = \boldsymbol{\sigma}_0^d - \boldsymbol{\sigma}_0$, with $\boldsymbol{\sigma}_0^d = 0$, ends up following an exponentially stable dynamics

$$\dot{\boldsymbol{\sigma}}_0 = -\boldsymbol{\Lambda}_0 \boldsymbol{\sigma}_0 . \tag{4.17}$$

4.3.1 Motion Distribution

In those tasks involving platform and arm movements (i.e., the task Jacobian $\boldsymbol{J}_0$ includes the platform and arm Jacobians) we can penalize the motion of the multirotor against the arm to account for their different motion capabilities. For example, this is the case of performing a visual servo task with an eye-in-hand configuration (4.13). To do so, we can define as in [9] a weighted norm of the whole velocity vector with

$$\|\dot{\rho}\|_W = \sqrt{\dot{\rho}^\top W \dot{\rho}}\,, \tag{4.18}$$

and use a weighted task Jacobian to solve for the weighted controls

$$\dot{\rho}_W = W^{-1/2}\big(J_0\, W^{-1/2}\big)^+\, \varepsilon = J_0^{\#}\, \varepsilon\,, \tag{4.19}$$

with ε as in (4.15) and

$$J_0^{\#} = W^{-1}\, J_0^\top \,\big(J_0\, W^{-1}\, J_0^\top\big)^{-1} \tag{4.20}$$

the weighted generalized Moore-Penrose pseudoinverse of the servo Jacobian. With this, large movements should be achieved by the multirotor whereas the precise movements should be devoted to the robotic arm due to its dexterity when the platform is close to the target. To achieve this behavior, we define a time-varying diagonal weight-matrix, as proposed in [28],

$$W(d) = \mathrm{diag}\big((1-\iota)\, I_4,\, \iota\, I_n\big), \tag{4.21}$$

with $n = 4 + m$ the whole UAM DoFs (4 for the multirotor and m for the arm) and

$$\iota(d) = \frac{1+\underline{\iota}}{2} + \frac{1-\underline{\iota}}{2}\tanh\left(2\pi\,\frac{d-\delta_W}{\Delta_W-\delta_W} - \pi\right), \tag{4.22}$$

where $\iota \in [\underline{\iota}, 1]$, and δ_W and Δ_W, $\Delta_W > \delta_W$, are the distance thresholds corresponding to $\iota \cong 1$ and $\iota \cong \underline{\iota}$, respectively. The blocks of W weight differently the velocity components of the arm and the multirotor by increasing the velocity of the multirotor when the distance to the target $d > \Delta_W$, while for distances $d < \delta_W$ the multirotor is slowed down and the arm is commanded to accommodate for the precise movements.

4.4 Hierarchical Task Priority Control (HTPC)

In case of a redundant robotic arm (i.e., $m > 2$), we can define several tasks and prioritize them depending on our objectives. One of the most used strategies to set a priority consist on a weighted sum of subtasks as in [28]. However, this technique can be problematic when the tasks are antagonistic. Instead, another solution is to adopt a hierarchical control law to prioritize the tasks and force the accomplishment of those more critical. In this section we present two similar approaches to hierarchical task composition. First, we describe the work presented in [29] which consists on a full least squares task solution. Then, we define a similar hierarchical control law, but in contrast to the previous solution, in this case we only require independence of the non controllable DoFs between tasks. This last control law was presented in [51].

4.4.1 HTPC Using Full Least Squares

Let us consider a secondary lower priority task $\boldsymbol{\sigma}_1 = \boldsymbol{f}_1(\boldsymbol{x})$ such that

$$\dot{\boldsymbol{\sigma}}_1 = \frac{\partial \boldsymbol{f}_1(\boldsymbol{x})}{\partial \boldsymbol{x}} \dot{\boldsymbol{x}} = \boldsymbol{J}_1 \dot{\boldsymbol{\rho}}_1 + \overline{\boldsymbol{J}}_1 \boldsymbol{\varpi} , \tag{4.23}$$

with $\dot{\boldsymbol{\rho}}_1 = \boldsymbol{J}_1^+ \left(\boldsymbol{\Lambda}_1 \widetilde{\boldsymbol{\sigma}}_1 - \overline{\boldsymbol{J}}_1 \boldsymbol{\varpi}\right)$. We can define a task composition strategy that minimizes secondary task velocity reconstruction only for those components in (4.23) that do not conflict with the primary task [1], namely

$$\dot{\boldsymbol{\rho}} = \boldsymbol{J}_0^+ \boldsymbol{\Lambda}_0 \widetilde{\boldsymbol{\sigma}}_0 + \left(\boldsymbol{J}_1 \boldsymbol{N}_0\right)^+ \boldsymbol{\Lambda}_1 \widetilde{\boldsymbol{\sigma}}_1 - \overline{\boldsymbol{J}}_{0|1} \boldsymbol{\varpi} , \tag{4.24}$$

where $\boldsymbol{N}_0 = \boldsymbol{I}_n - \boldsymbol{J}_0^+ \boldsymbol{J}_0$ is the projector onto the null space of $\boldsymbol{J}_0$, $\boldsymbol{\Lambda}_1$ is a square positive definite gain matrix, $\boldsymbol{J}_1$ the Jacobian matrix of the second subtask, which is assumed to be full-rank, and

$$\overline{\boldsymbol{J}}_{0|1} = \left(\boldsymbol{J}_1 \boldsymbol{N}_0\right)^+ \overline{\boldsymbol{J}}_1 + \left(\boldsymbol{I} - (\boldsymbol{J}_1 \boldsymbol{N}_0)^+ \boldsymbol{J}_1\right) \overline{\boldsymbol{J}}_{0|0}, \tag{4.25}$$

where $\boldsymbol{I}$ is the identity matrix and $\overline{\boldsymbol{J}}_{0|0} = \boldsymbol{J}_0^+ \overline{\boldsymbol{J}}_0$.

The Jacobian matrix $\overline{\boldsymbol{J}}_{0|1}$ allows the compensation of the variation of $\boldsymbol{\varpi}$. Notice that the matrix $\boldsymbol{J}_1 \boldsymbol{N}_0$ is full-rank only if the two tasks are *orthogonal* (i.e., $\boldsymbol{J}_1 \boldsymbol{J}_0^+ = \boldsymbol{0}$) or *independent* (i.e., not orthogonal and $\text{rank}(\boldsymbol{J}_0^+) + \text{rank}(\boldsymbol{J}_1^+) = \text{rank}\left([\boldsymbol{J}_0^+ \; \boldsymbol{J}_1^+]\right)$). Refer to [2] for more details. Substituting (4.24) into (4.14) and by noticing that $\boldsymbol{N}_0$ is idempotent and Hermitian, hence $(\boldsymbol{J}_1 \boldsymbol{N}_0)^+ = \boldsymbol{N}_0 (\boldsymbol{J}_1 \boldsymbol{N}_0)^+$, the dynamics of the main task (4.17) is again achieved and so the exponential stability is proven.

To study the behavior of the secondary task $\boldsymbol{\sigma}_1$ we can substitute (4.24) in (4.23), considering the task errors $\widetilde{\boldsymbol{\sigma}}_i = \boldsymbol{\sigma}_i^d - \boldsymbol{\sigma}_i$ with $\boldsymbol{\sigma}_i^d = 0$ and $i = 0..1$, and by assuming that the tasks are at least independent, the following dynamics is achieved

$$\begin{aligned} \dot{\boldsymbol{\sigma}}_1 =& - \boldsymbol{J}_1 \boldsymbol{J}_0^+ \boldsymbol{\Lambda}_0 \boldsymbol{\sigma}_0 - \boldsymbol{J}_1 (\boldsymbol{J}_1 \boldsymbol{N}_0)^+ \boldsymbol{\Lambda}_1 \boldsymbol{\sigma}_1 \\ &+ \left(\boldsymbol{J}_1 \boldsymbol{J}_0^+ \overline{\boldsymbol{J}}_0 + \boldsymbol{J}_1 (\boldsymbol{J}_1 \boldsymbol{N}_0)^+ (\overline{\boldsymbol{J}}_1 - \boldsymbol{J}_1 \boldsymbol{J}_0^+ \overline{\boldsymbol{J}}_0) - \overline{\boldsymbol{J}}_1\right) \boldsymbol{\varpi} \\ =& - \boldsymbol{J}_1 \boldsymbol{J}_0^+ \boldsymbol{\Lambda}_0 \boldsymbol{\sigma}_0 - \boldsymbol{\Lambda}_1 \boldsymbol{\sigma}_1, \end{aligned} \tag{4.26}$$

where we used the property $\boldsymbol{J}_1 (\boldsymbol{J}_1 \boldsymbol{N}_0)^+ = \boldsymbol{I}$.

Finally the dynamics of the system can be written as

$$\begin{bmatrix} \dot{\boldsymbol{\sigma}}_0 \\ \dot{\boldsymbol{\sigma}}_1 \end{bmatrix} = \begin{bmatrix} -\boldsymbol{\Lambda}_0 & \boldsymbol{0} \\ -\boldsymbol{J}_1 \boldsymbol{J}_0^+ \boldsymbol{\Lambda}_0 & -\boldsymbol{\Lambda}_1 \end{bmatrix} \begin{bmatrix} \boldsymbol{\sigma}_0 \\ \boldsymbol{\sigma}_1 \end{bmatrix}, \tag{4.27}$$

that is characterized by a Hurwitz matrix, hence the exponential stability of the system is guaranteed. Moreover, we can notice the term $-\boldsymbol{J}_1 \boldsymbol{J}_0^+ \boldsymbol{\Lambda}_0$ that couples the effect of the main task on the secondary task. In case of orthogonal tasks $\boldsymbol{J}_1 \boldsymbol{J}_0^+ = \boldsymbol{0}$

and $\dot{\boldsymbol{\sigma}}_1 = -\boldsymbol{\Lambda}_1 \boldsymbol{\sigma}_1$, and the behavior of the main and that of the secondary tasks are decoupled.

By generalizing (4.24) to the case of η prioritized subtasks, we can formulate the general case with

$$\dot{\boldsymbol{\rho}} = \boldsymbol{J}_0^+ \boldsymbol{\Lambda}_0 \widetilde{\boldsymbol{\sigma}}_0 + \sum_{i=1}^{\eta} (\boldsymbol{J}_i \boldsymbol{N}_{0|\ldots|i-1})^+ \boldsymbol{\Lambda}_i \widetilde{\boldsymbol{\sigma}}_i - \overline{\boldsymbol{J}}_{0|\ldots|\eta} \boldsymbol{\varpi}, \tag{4.28}$$

where the recursively-defined compensating matrix is

$$\overline{\boldsymbol{J}}_{0|\ldots|\eta} = (\boldsymbol{J}_\eta \boldsymbol{N}_{0|\ldots|\eta-1})^+ \overline{\boldsymbol{J}}_\eta + \left(\boldsymbol{I} - (\boldsymbol{J}_\eta \boldsymbol{N}_{0|\ldots|\eta-1})^+ \boldsymbol{J}_\eta\right) \overline{\boldsymbol{J}}_{0|\ldots|\eta-1}, \tag{4.29}$$

with $\boldsymbol{N}_{0|\ldots|i}$ the projector onto the null space of the *augmented Jacobian* $\boldsymbol{J}_{0|\ldots|i}$ of the ith subtask, with $i = 0, \ldots, \eta - 1$, which are defined as

$$\boldsymbol{J}_{0|\ldots|i} = \left[\boldsymbol{J}_0^\top \cdots \boldsymbol{J}_i^\top\right]^\top \tag{4.30a}$$

$$\boldsymbol{N}_{0|\ldots|i} = (\boldsymbol{I} - \boldsymbol{J}_{0|\ldots|i}^+ \boldsymbol{J}_{0|\ldots|i}). \tag{4.30b}$$

The previous stability analysis can be straightforwardly extended to the general case of η subtasks.

The hierarchical formulation (4.28) guarantees that the execution of all the higher-priority tasks from 0 (main task) to $i - 1$ will not be affected by the ith subtask and by the variation of the uncontrolled state variables. In other words, the execution of the ith task is subordinated to the execution of the higher priority tasks present in the task stack, i.e., it will be fulfilled only if suitable and enough DoFs are available, while the complete fulfillment of the main task, instead, is always guaranteed. However, with this new formulation for all reciprocally annihilating or independent tasks, a fully decoupling of the error dynamics is guaranteed.

4.4.2 HTPC Decoupling Algorithmic Task Singularities

Considering again a secondary lower priority task $\boldsymbol{\sigma}_1 = \boldsymbol{f}_1(\boldsymbol{x})$ as in (4.23), we can define a task composition strategy that minimizes secondary task velocity reconstruction only for those components in (4.23) that do not conflict with the primary task. However, in this case we formulate the hierarchical composition with

$$\dot{\boldsymbol{\rho}} = \boldsymbol{J}_0^+ \boldsymbol{\Lambda}_0 \widetilde{\boldsymbol{\sigma}}_0 + \boldsymbol{N}_0 \boldsymbol{J}_1^+ \boldsymbol{\Lambda}_1 \widetilde{\boldsymbol{\sigma}}_1 - \overline{\boldsymbol{J}}_{0|1} \boldsymbol{\varpi}, \tag{4.31}$$

where $\boldsymbol{N}_0 = (\boldsymbol{I}_n - \boldsymbol{J}_0^+ \boldsymbol{J}_0)$ is the null space projector of the primary task and

$$\overline{\boldsymbol{J}}_{0|1} = \boldsymbol{J}_0^+ \overline{\boldsymbol{J}}_0 + \boldsymbol{N}_0 \boldsymbol{J}_1^+ \overline{\boldsymbol{J}}_1 \tag{4.32}$$

is the Jacobian matrix that allows for the compensation of the variation of the uncontrollable states $\boldsymbol{\varpi}$.

This strategy, in contrast to the more restrictive one we presented in (4.24) might achieve larger constraint-task reconstruction errors than the full least squares secondary task solution of the previous section but with the advantage that algorithmic singularities arising from conflicting tasks are decoupled from the singularities of the secondary tasks.

It remains to show that this task composition strategy guarantees stability of the overall system. Substituting (4.31) into (4.23), and considering a task error $\widetilde{\boldsymbol{\sigma}}_1 = \boldsymbol{\sigma}_1^d - \boldsymbol{\sigma}_1$, the following dynamics for the secondary task is achieved

$$\dot{\boldsymbol{\sigma}}_1 = -\boldsymbol{J}_1\boldsymbol{J}_0^+\boldsymbol{\Lambda}_0\boldsymbol{\sigma}_0 - \boldsymbol{\Lambda}_1\boldsymbol{\sigma}_1 + (\boldsymbol{J}_1\boldsymbol{J}_0^+\overline{\boldsymbol{J}}_0)\boldsymbol{\varpi}\,, \tag{4.33}$$

where we used the property $\boldsymbol{J}_1\boldsymbol{N}_0\boldsymbol{J}_1^+ = \boldsymbol{I}$. Notice how exponential stability of this last expression can only be guaranteed when the tasks are independent for the uncontrollable states $\boldsymbol{\varpi}$, i.e., $\boldsymbol{J}_1\boldsymbol{J}_0^+\overline{\boldsymbol{J}}_0 = \boldsymbol{0}$, hence

$$\begin{bmatrix}\dot{\boldsymbol{\sigma}}_0\\ \dot{\boldsymbol{\sigma}}_1\end{bmatrix} = \begin{bmatrix}-\boldsymbol{\Lambda}_0 & \boldsymbol{0}\\ -\boldsymbol{J}_1\boldsymbol{J}_0^+\boldsymbol{\Lambda}_0 & -\boldsymbol{\Lambda}_1\end{bmatrix}\begin{bmatrix}\boldsymbol{\sigma}_0\\ \boldsymbol{\sigma}_1\end{bmatrix}, \tag{4.34}$$

which is a less stringent condition than whole task orthogonality ($\boldsymbol{J}_1\boldsymbol{J}_0^+ = \boldsymbol{0}$) or independence that was needed in (4.24) to compute the pseudo-inverse $(\boldsymbol{J}_1\boldsymbol{N}_0)^+$.

The addition of more tasks in cascade is possible as long as there exist remaining DoFs from the concatenation of tasks higher up in the hierarchy. The generalization of (4.31) to the case of η prioritized subtasks is

$$\dot{\boldsymbol{\rho}} = \boldsymbol{J}_0^+\boldsymbol{\Lambda}_0\widetilde{\boldsymbol{\sigma}}_0 + \sum_{i=1}^{\eta}\boldsymbol{N}_{0|\ldots|i-1}\boldsymbol{J}_i^+\boldsymbol{\Lambda}_i\widetilde{\boldsymbol{\sigma}}_i - \overline{\boldsymbol{J}}_{0|\ldots|\eta}\,\boldsymbol{\varpi} \tag{4.35}$$

with the recursively-defined compensating matrix

$$\overline{\boldsymbol{J}}_{0|\ldots|\eta} = \boldsymbol{N}_{0|\ldots|i-1}\boldsymbol{J}_i^+\overline{\boldsymbol{J}}_i + (\boldsymbol{I} - \boldsymbol{N}_{0|\ldots|i-1}^+\boldsymbol{J}_i^+\overline{\boldsymbol{J}}_i)\overline{\boldsymbol{J}}_{0|\ldots|i-1}\,, \tag{4.36}$$

where $\boldsymbol{N}_{0|\ldots|i}$ is the projector onto the null space of the augmented Jacobian $\boldsymbol{J}_{0|\ldots|i}$ for the i-th subtask, with $i = 0, \ldots, \eta - 1$, defined as in (4.30).

4.4.3 Dynamic Change of Task Priorities

The task composition and priority can be modified at runtime as needed, i.e., by activating or deactivating subtasks as well as by changing the priority order of the current active tasks already present in the task stack. However, in order to avoid discontinuity of the control input, a smooth transition between different task stacks has to be considered. This goal can be achieved by adopting a time-vanishing *smoothing term* when a new task stack is activated, as explained in the following.

Without loss of generality, we suppose that the transition phase starts at $t = 0$, i.e., the rth task stack has to be deactivated and substituted by the new one $(r + 1)$th. During the transition the velocity command is computed with

$$\dot{\boldsymbol{\rho}}(t) = \dot{\boldsymbol{\rho}}_{r+1}(t) + e^{-\frac{t}{t_c}} \left(\dot{\boldsymbol{\rho}}_r(0) - \dot{\boldsymbol{\rho}}_{r+1}(0) \right), \tag{4.37}$$

where t_c is a time constant determining the transition phase duration, and $\dot{\boldsymbol{\rho}}_k$ is the velocity command corresponding to the kth task stack. When t becomes sufficiently greater than t_c, the rth task stack is fully removed and a new transition can start. Notice that the smoothing term $e^{-\frac{t}{t_c}} \left(\dot{\boldsymbol{\rho}}_r(0) - \dot{\boldsymbol{\rho}}_{r+1}(0) \right)$ is bounded and exponentially vanishing, hence it will not affect the stability of the proposed control law (the assumptions of Lemma 9.1 in [20] can be easily verified if all the Jacobian matrices are full rank, hence the system is globally exponentially stable). The time constant t_c has to be smaller than the inverse of the maximum eigenvalue of the gain matrices $\boldsymbol{\Lambda}_i$ to ensure a short transient time response in comparison with the nullifying time of the task errors $\widetilde{\boldsymbol{\sigma}}_i$.

4.5 Optimization-Based Trajectory Generation

The goal of this Section is to describe an optimization method to generate feasible trajectories for all the joints of a multirotor-arm system, which has been published in [49]. In contrast to the hierarchical control laws presented in Sect. 4.4.1 and 4.4.2, here we take advantage of a quadratic programming approach (QP) to optimize robot joint commands to accomplish the tasks. Although in this case the task priorities are assigned using a weighting strategy, the numerical optimization problem allows us to set bounds and constraints. Hence, we can set the critical objectives (e.g., collision avoidance) as a constraint for all other tasks.

Let us consider a UAM model as in previous Sections, composed by a multirotor platform with 4 DoFs and an arm with m joints (i.e., any of the settings presented in Fig. 4.1). Moreover, we consider that UAMs are not meant for acrobatic maneuvers. Hence, we have not included the platform tilt in the trajectory generation algorithm (roll and pitch angles will be assumed negligible in our analysis). Thus, we use here the reduced system state vector $\boldsymbol{\rho}$ and the UAM DoFs $\dot{\boldsymbol{\rho}}$ from (4.4) and (4.5), respectively.

In the first part of this Section, we will assume that the inner control loop of the system can perfectly track the computed references. However, this hypothesis will be removed in Sect. 4.5.4 and its implications discussed.

4.5.1 Optimization Principles

The goal of a trajectory generation algorithm is to command the robot DoFs to accomplish some given tasks while satisfying the system constraints. These tasks and constraints can be generically expressed by

$$\min f_i(\boldsymbol{\rho}, \dot{\boldsymbol{\rho}}, t) \quad \text{and} \quad f_j(\boldsymbol{\rho}, \dot{\boldsymbol{\rho}}, t) \leq 0\,, \tag{4.38}$$

where $\min f_i(\boldsymbol{\rho}, \dot{\boldsymbol{\rho}}, t)$ represents the ith generic task and $f_j(\boldsymbol{\rho}, \dot{\boldsymbol{\rho}}, t) \leq 0$ stands for the jth constraint. For example, a trajectory tracking task with the arm end effector can be expressed as the minimization of the tracking error norm, $f(\boldsymbol{\rho}, t) = ||\boldsymbol{p}_T^d(t) - \boldsymbol{p}_T(t)||$, where $\boldsymbol{p}_T^d(t)$ and $\boldsymbol{p}_T(t)$ are the desired and actual end effector positions, respectively (the subscript T indicates the tool attached to the end effector to ease the reading).

The key idea of this approach is to assign desired dynamics to $f_i(\boldsymbol{\rho}, \dot{\boldsymbol{\rho}}, t)$ and $f_j(\boldsymbol{\rho}, \dot{\boldsymbol{\rho}}, t)$. In fact, by using the Gronwall inequality as in [5], a constraint expressed as $f \leq 0$ is satisfied if

$$\dot{f} \leq -\lambda_1 f, \tag{4.39}$$

where λ_1 is a positive scalar gain. Notice that it is just a sufficient condition, since the inequality of (4.39) is more restrictive than the original one.

By applying iteratively the Gronwall inequality on $\dot{f} + \lambda_1 f \leq 0$ from (4.39), we have

$$\ddot{f} \leq -\lambda_2 \left(\dot{f} + \lambda_1 f\right) - \lambda_1 \dot{f}, \tag{4.40}$$

where λ_2 is a positive scalar gain. Parameters λ_1 and λ_2 assign the maximum convergence dynamics towards the constraint. Similarly to the constraints, a task expressed by $\min f$, where $f \geq 0$, can be formulated as

$$\min ||\ddot{f} + (\lambda_2 + \lambda_1)\, \dot{f} + \lambda_2 \lambda_1 f||^2. \tag{4.41}$$

Notice that if the cost function in (4.41) is always kept at its lower bound, the function f converges to 0 with a dynamics assigned by eigenvalues λ_1 and λ_2.

This approach is useful to obtain constraints and cost functions (i.e., tasks) in the acceleration domain, when the original ones are expressed in the position domain. In fact, considering the constraint $f_j(\boldsymbol{\rho}) \leq 0$ depending only on robot joint values, and applying this approach, we end up with

$$\boldsymbol{A}_j \ddot{\boldsymbol{\rho}} \leq u_{Aj}, \tag{4.42}$$

where

$$\boldsymbol{A}_j = \frac{\partial f_j}{\partial \boldsymbol{\rho}}, \tag{4.43a}$$

$$u_{Aj} = -\lambda_2\lambda_1 f_j - \left(\dot{\boldsymbol{\rho}}^\top \frac{\partial^2 f_j}{\partial \boldsymbol{\rho}^2} + (\lambda_1 + \lambda_2)\frac{\partial f_j}{\partial \boldsymbol{\rho}}\right)\dot{\boldsymbol{\rho}}\,. \tag{4.43b}$$

On the other hand, when a constraint also depends on joint velocities $f_j(\boldsymbol{\rho}, \dot{\boldsymbol{\rho}}) \leq 0$) the Gronwall inequality should be applied only once. While the constraint expression is the same as in (4.42), in this case the terms $\boldsymbol{A}_j$ and u_{Aj} are computed with

$$\boldsymbol{A}_j = \frac{\partial f_j}{\partial \dot{\boldsymbol{\rho}}}, \quad u_{Aj} = -\lambda_2 f_j - \frac{\partial f_j}{\partial \boldsymbol{\rho}}\dot{\boldsymbol{\rho}}\,. \tag{4.44}$$

Similarly to the linear formulation obtained for the constraints (4.42), the cost functions can be defined as

$$\min ||\dot{\boldsymbol{J}}_i \ddot{\boldsymbol{\rho}} - \boldsymbol{b}_i||^2, \tag{4.45}$$

where, similarly to the constraints, the computations of $\dot{\boldsymbol{J}}_i$ and $\boldsymbol{b}_i$ are straightforward as in (4.44).

Notice that, so far the analysis has been performed for scalar functions f_i, such that $\dot{\boldsymbol{J}}_i$ results in a row vector, however it can be easily extended to multidimensional tasks $\boldsymbol{f}_i$ and Jacobian matrices $\dot{\boldsymbol{J}}_i$.

4.5.2 Quadratic Problem Formulation

The formulation of the optimization problem is quite straightforward. The cost function in (4.45) results in the quadratic form

$$\min_{\boldsymbol{a}} \|\dot{\boldsymbol{J}}_i \boldsymbol{a} - \boldsymbol{b}_i\|^2 = \min_{\boldsymbol{a}} \left(\boldsymbol{a}^\top \dot{\boldsymbol{J}}_i^\top \dot{\boldsymbol{J}}_i \boldsymbol{a} - 2\boldsymbol{b}_i^\top \dot{\boldsymbol{J}}_i \boldsymbol{a}\right), \tag{4.46}$$

where the regressor variable $\ddot{\boldsymbol{\rho}}$ has been replaced by $\boldsymbol{a}$ for simplicity.

As we want to minimize different objective functions (i.e., tasks), two different scenarios are possible. In the case that a strict hierarchy between tasks is required, a hierarchical solver has to be used (e.g., [12]). Alternatively, we can consider the case of partial hierarchy setting critical tasks as constraints, and adopting a weighted sum of the other n_T objective functions with

$$\begin{aligned} \min_{\boldsymbol{a}} \sum_{i=1}^{n_T} \sigma_i &= \min_{\boldsymbol{a}} \sum_{i=1}^{n_T} \frac{w_i}{h_i}\left(\boldsymbol{a}^\top \dot{\boldsymbol{J}}_i^\top \dot{\boldsymbol{J}}_i \boldsymbol{a} - 2\boldsymbol{b}_i^\top \dot{\boldsymbol{J}}_i \boldsymbol{a}\right) \\ &= \min_{\boldsymbol{a}} \sum_{i=1}^{n_T} \frac{w_i}{h_i}\left(\boldsymbol{a}^\top \boldsymbol{H}_i \boldsymbol{a} + \boldsymbol{m}_i^\top \boldsymbol{a}\right), \end{aligned} \tag{4.47}$$

where w_i and h_i are weights and a normalization factor, respectively. When a weighted sum is employed, it is very important to normalize the objective functions, in order to effectively set the desired weights w_i. Thus, we chose the factors h_i equal to the spectral norm of $\boldsymbol{H}_i$, which is equal to the square root of the largest eigenvalue of the product matrix $\boldsymbol{H}_i^\top \boldsymbol{H}_i$. Notice that this spectral norm of $\boldsymbol{H}_i$ equals the square of the spectral norm of $\dot{\boldsymbol{J}}_i$ when the objective function has the form as in (4.46). Moreover, in order to effectively use the normalization factor h_i, it is beneficial to split the tasks that are not dimensionally coherent. For instance, the end effector error is composed of translational and rotational parts. Then, computing two different factors h_i improves the effectiveness of the normalization.

With this formulation, the norms of joint velocities and accelerations define two useful cost functions. They assure the convexity of the problem and allow the distribution of the motion on the different joints a priori, by assigning different weights to each joint. When these two cost functions are used together, the weights on joint velocities should be larger by a factor 5–10 than the corresponding weights on joint accelerations, in order to obtain a coherent behavior.

Finally, the complete optimization system combines the cost functions and constraints, obtaining a quadratic problem with linear constraints defined as

$$\begin{aligned} \min_{\boldsymbol{a}} \boldsymbol{\sigma} &= \min_{\boldsymbol{a}} \left(\frac{1}{2} \boldsymbol{a}^\top \boldsymbol{H} \boldsymbol{a} + \boldsymbol{m}^\top \boldsymbol{a} \right) \\ s.t. \quad & \boldsymbol{l}_b \leq \boldsymbol{a} \leq \boldsymbol{u}_b \\ & \boldsymbol{l}_A \leq \boldsymbol{A}\boldsymbol{a} \leq \boldsymbol{u}_A \, , \end{aligned} \tag{4.48}$$

where $\boldsymbol{\sigma} = \sum \sigma_i$, $\boldsymbol{H} = \sum \frac{w_i}{h_i} \boldsymbol{H}_i$, $\boldsymbol{m} = \sum \frac{w_i}{h_i} \boldsymbol{m}_i$ and $\boldsymbol{A} = \left[\boldsymbol{A}_1^\top \ \dots \ \boldsymbol{A}_{n_C}^\top\right]^\top$ with n_C the number of constraints. $\boldsymbol{l}_b$, $\boldsymbol{u}_b$, $\boldsymbol{l}_A$ and $\boldsymbol{u}_A$ assign lower ($\boldsymbol{l}_x$) and upper ($\boldsymbol{u}_x$) bounds on acceleration and constraints respectively.

Then, we obtain a solution $\boldsymbol{a}_0$ of the system in (4.48) for every time step k, thus the position and velocity references, (4.4) and (4.5), can be updated, for example, with an Euler integration as

$$\boldsymbol{\rho}(k) = \boldsymbol{\rho}(k-1) + \dot{\boldsymbol{\rho}}(k-1)\,\delta t + \boldsymbol{a}_0 \frac{\delta t^2}{2} \tag{4.49a}$$

$$\dot{\boldsymbol{\rho}}(k) = \dot{\boldsymbol{\rho}}(k-1) + \boldsymbol{a}_0\,\delta t . \tag{4.49b}$$

where δt is the time step differential.

It is important to remark that the trajectory generation algorithm can either be computed off-line or on-line. As the quadratic optimization method can be efficiently solved by state of the art algorithms, it is particularly suitable to be computed on-line even in limited computational units. On-line iteration allows to dynamically change the optimization parameters during specific phases of a mission. It can be particularly useful to implement different redundancy solution strategies, by changing cost function weights, depending on the mission phase or external triggers. This behavior is shown in the experiments Section for a specific UAM mission.

4.5.3 Position, Velocity and Acceleration Bounds

Using numerical optimization techniques allows us to specify robot constraints and tasks, as the ones presented in Sect. 4.6, specific for aerial manipulators. However, to reach an optimum solution we have to include in the optimization problem the position, velocity and acceleration limits of the UAM joints. For instance, the quadrotor DoF for height has an obvious lower bound (i.e., the ground), while arm joint bounds are given by the physical arm structure. These problem bounds have to be reported in acceleration form with the strategy presented in Sect. 4.5.1.

Given an upper bound on the i-th joint position, namely $\overline{q}_i$, the Gronwall inequality can be applied twice as in (4.40), obtaining the following position bound in acceleration form

$$\ddot{q}_i \leq -(\lambda_1 + \lambda_2)\,\dot{q}_i - \lambda_1\lambda_2\left(q_i - \overline{q}_i\right) . \tag{4.50}$$

On the other hand, when an upper bound $\dot{\overline{q}}_i$ is given on the i-th joint velocity, the Gronwall inequality has to be applied just once as in (4.39), resulting in the following velocity bound

$$\ddot{q}_i \leq -\lambda_1\left(\dot{q}_i - \dot{\overline{q}}_i\right) . \tag{4.51}$$

The vectors of lower and upper bounds, $\boldsymbol{l}_b$ and $\boldsymbol{u}_b$ in (4.48), are formed considering the most stringent conditions between position (4.50), velocity (4.51) and acceleration bounds, for each joint.

4.5.4 Interface with the Control Algorithm

So far in this Sect. 4.5 we have assumed a perfect tracking of the generated reference trajectory by the inner controller. With a real system, the performance of this closed control loop is not ideal and dynamics between desired ($\boldsymbol{\rho}^d$) and actual ($\boldsymbol{\rho}$) values of the robot DoFs are introduced. As a consequence, the parameters λ_1 and λ_2 should be set low enough with respect the control bandwidth such that the time scale separation principle holds, and the reference can be tracked. Alternatively, one can include a simplified model of the closed loop system in the dynamics considered by the optimization. Such dynamics analysis is out of the book scope.

In addition, if the actual measures are fed into the optimization algorithm (feedback scheme), feasibility and stability issues can arise. In fact, using this feedback requires techniques of robust optimization, as in [57], to prevent unfeasible points. Moreover, the closed loop control can affect the stability of the system. The nonlinearity of the optimization method makes it really difficult to find stability bounds and conditions.

For the previous reasons, in the rest of the Section the trajectory generation algorithm, as it is common in all robotic trajectory generation systems, is computed without any feedback of joint measures, and parameters λ_1 and λ_2 are chosen according to the inner control bandwidth and time step δt.

4.6 Tasks Definitions

In this Section we present several elementary tasks to be carried out during UAM missions, in order to accomplish a given objective or to preserve stability. Gripper operations like grasping are not considered here. These tasks are formulated in order to be used either by the hierarchical approaches presented in Sect. 4.4 or by the quadratic programing approach of Sect. 4.5. For the latter, we do not mention here the weighting factors because they depend on particular mission phases, as explained in Sect. 4.7.3. Moreover, the expressions of the task Jacobian derivatives $\dot{\boldsymbol{J}}_i$ and the terms $\boldsymbol{b}_i$ are neither provided but can be obtained using (4.45), which in turn is based on (4.44).

4.6.1 Collision Avoidance

The most important task during a mission is to preserve flight safety. When a rotor operates near an obstacle, different aerodynamic effects are revealed, such as the so called "ground" or "ceiling" effects, that can lead to an accident. Hence, to avoid them, we propose a task with the highest priority to maintain a certain distance to obstacles by defining a safety sphere around the flying platform, as shown in Fig. 4.2a, and comparing its radius (r_o) with the Euclidean distance to the obstacle ($\boldsymbol{d}_o$). This distance can also be computed with respect to a different moving UAM elements. For example, if we set the center of the sphere at the arm end effector, the same task formulation can be used to avoid self-collisions by computing the distance $\boldsymbol{d}_o$ with respect to own robot parts, e.g., legs in Fig. 4.2b.

Then, our task function to minimize becomes

$$\sigma_o = r_o - ||\boldsymbol{d}_o|| \,, \tag{4.52}$$

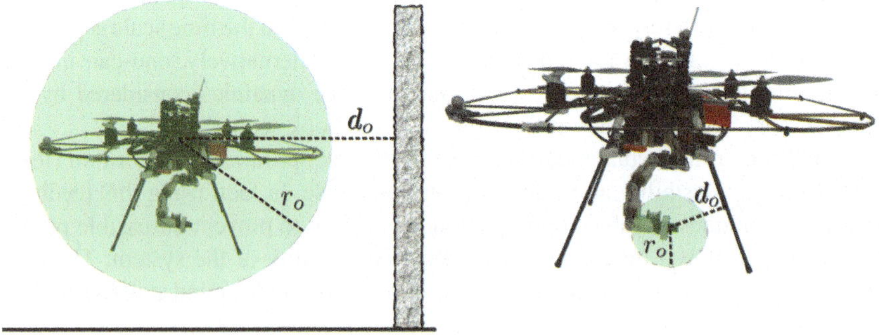

(a) Obstacle avoidance. (b) Self-collision avoidance.

Fig. 4.2 Collision avoidance setting an inflation radius around a specific part of the UAM. In frame **a** this radius is set to avoid avoid obstacles, whereas in frame **b** is set to avoid self-collisions

The desired task variable is $\sigma_o^d = 0$ (i.e., $\widetilde{\sigma}_o = -\lambda_o \sigma_o$ with λ_o a suitable positive gain), while the corresponding task Jacobian is

$$\boldsymbol{J}_o = -2\boldsymbol{d}_o^\top \boldsymbol{S}_p \boldsymbol{J}_{R_O}, \tag{4.53}$$

where $\boldsymbol{S}_p = \begin{bmatrix} \boldsymbol{I}_3 & \boldsymbol{0}_3 \end{bmatrix}$ is a matrix to select the first three rows of the Jacobian $\boldsymbol{J}_{R_O}$, which is a $6 \times (m+4)$ Jacobian matrix transforming controllable UAM DoFs values to a velocity vector, and it depends on where do we set the center of the sphere. In case of avoiding collisions with the platform, and setting the sphere center in the platform body frame b, we have

$$\boldsymbol{J}_{R_O} = \begin{bmatrix} \boldsymbol{I}_3 & \boldsymbol{0}_{3\times(m+1)} \\ \boldsymbol{0}_3 & \boldsymbol{0}_{3\times(m+1)} \end{bmatrix}. \tag{4.54}$$

The Jacobian matrix of the uncontrolled state variables is $\overline{\boldsymbol{J}}_o = -2\boldsymbol{d}_o^\top \boldsymbol{S}_p \overline{\boldsymbol{J}}_{R_O} = \boldsymbol{0}_{1\times 2}$.

If we set the task to avoid self-collisions with the arm end effector, the Jacobian $\boldsymbol{J}_{R_O}$ corresponds to the controllable part of the robot Jacobian presented in (4.13) (i.e., columns corresponding to controllable DoFs). Similarly we can formulate $\overline{\boldsymbol{J}}_{R_P}$ with the non controllable columns of (4.13). Notice how with an *onboard-eye-to-hand* configuration we have to change the subscripts c by t in (4.13).

When using a hierarchical control law, if we define the collision avoidance task with the highest priority we have to deactivate its effects if no potential collision is detected (i.e., invasion of the sphere). As the hierarchy is based on the dimension of the null space of the higher priority task Jacobians, we can define a generalized pseudoinverse similarly to (4.20) with a diagonal activation matrix, thus

$$\boldsymbol{J}_o^{\#} = \boldsymbol{H}^{-1} \boldsymbol{J}_o^\top \left(\boldsymbol{J}_o \boldsymbol{H}^{-1} \boldsymbol{J}_o^\top \right)^{-1}. \tag{4.55}$$

With this activation matrix, we can prevent potential collisions by cancelling the motion of the element (e.g., the flying platform or the end effector) only in those directions susceptible to collide. This is usually called joint clamping. The elements in $\boldsymbol{H}$ are

$$\boldsymbol{H} = \mathrm{diag}\big(h_x, h_y, h_z, \boldsymbol{0}_{1\times(m+1)}\big), \tag{4.56}$$

and will block only those multirotor DoFs where the inflation radius is not respected defining

$$h_i = \begin{cases} 1, & \text{if } d_{oi} < (r_o - \|\boldsymbol{d}_{oj}\|) \forall j \neq i. \\ 0, & \text{otherwise.} \end{cases} \tag{4.57}$$

This control law clamps any motion that violates the minimum distance to the obstacle.

When using an optimization technique, such as the quadratic programming solution presented in Sect. 4.5, this collision avoidance task can be set as a constraint for the optimization problem, in order to prioritize it over the other tasks. As commonly done in numerical optimization problems, we can define the task function using the square norm of the distance to obstacles, becoming the inequality

$$\boldsymbol{d}_o^\top \boldsymbol{d}_o \geq r_o^2 \,, \tag{4.58}$$

resulting that $\boldsymbol{A}_o = 2\boldsymbol{d}_o^\top \boldsymbol{S}_p \boldsymbol{J}_R$ in (4.42), where with the definition of the optimization technique in Sect. 4.5, $\boldsymbol{J}_R$ is the full Jacobian of the UAM depending once again on the place we set the center of the sphere. For example, choosing it in the quadrotor body frame b, we have

$$\boldsymbol{J}_R = \begin{bmatrix} \boldsymbol{I}_3 & \boldsymbol{0}_{3\times(m+3)} \\ \boldsymbol{0}_3 & \boldsymbol{0}_{3\times(m+3)} \end{bmatrix} \,, \tag{4.59}$$

whereas for end effector self-collision avoidance it is directly (4.13).

4.6.2 Onboard-Eye-to-Hand: Global End Effector Tracking

The main interaction task with UAMs, in general, will be executed by the arm end effector, thus it is important to be able to track a desired end effector trajectory. To do so, here we present an end effector tracking task (positioning and orientation) with respect to a global frame, and considering the *onboard-eye-to-hand* configuration from Fig. 4.1. Notice how in the case of obtaining the current end effector pose by combining the visual information from the camera mounted in the platform and the arm kinematics, this task corresponds to a pose-based visual servo as seen in Sect. 3.4.

Considering a desired 3D position for the end effector expressed in the world frame w, we can compute its positioning error $\boldsymbol{e}_p$, and define the task function as its square norm, yielding

$$\sigma_p = \boldsymbol{e}_p^\top \boldsymbol{e}_p \,, \tag{4.60}$$

where the desired task variable is $\sigma_p^d = 0$ (i.e., $\widetilde{\sigma}_p = -\lambda_p \sigma_p$ with λ_p a positive definite gain).

The corresponding task Jacobian matrix is

$$\boldsymbol{J}_p = 2\boldsymbol{e}_p^\top \boldsymbol{J}_{R\,p}, \tag{4.61}$$

where $\boldsymbol{J}_{R\,p}$ corresponds to the robot Jacobian, consisting on the first three rows and the columns related to UAMs DoFs from $\boldsymbol{J}_R$ in (4.12). Similarly, the task Jacobian matrix of the uncontrolled state variables is

$$\overline{\boldsymbol{J}}_p = 2\boldsymbol{e}_p^\top \overline{\boldsymbol{J}_{R\,p}}, \tag{4.62}$$

where $\overline{\boldsymbol{J}_{R\,p}}$ is the Jacobian considering the first three rows and columns of the non controllable DoFs of $\boldsymbol{J}_R$ in (4.12).

With the proposed choice of σ_p, only one DoF is required to execute this subtask, because only the norm of $\boldsymbol{e}_p$ will be nullified, i.e., the motion of the gripper during the transient is constrained on a sphere of radius $\|\boldsymbol{e}_p\|$. However, the corresponding task Jacobian $\boldsymbol{J}_p$ becomes singular when $\boldsymbol{e}_p \rightarrow \boldsymbol{0}$. Nevertheless, in the task composition the generalized-inverse $\boldsymbol{J}_p^+$ is multiplied by σ_p. Hence, if $\boldsymbol{J}_p$ is full-rank, its determinant goes to zero only linearly when $\boldsymbol{e}_p \rightarrow \boldsymbol{0}$, but σ_p goes to zero squarely.

Regarding the end effector orientation error, we can use any of the error forms presented in Sect. 3.4. For example with the error computed using quaternions from (3.8), namely $\boldsymbol{e}_\phi$, we can define the task function as

$$\sigma_\phi = \boldsymbol{e}_\phi^\top \boldsymbol{e}_\phi, \tag{4.63}$$

with the desired task variable $\sigma_\phi^d = 0$ (i.e., $\widetilde{\sigma}_\phi = -\sigma_\phi$ with λ_ϕ a positive definite gain), while the corresponding task Jacobian matrix is

$$\boldsymbol{J}_\phi = 2\boldsymbol{e}_\phi^\top \boldsymbol{J}_{R\phi}, \tag{4.64}$$

where $\boldsymbol{J}_{R\phi}$ is now obtained from (4.12) just as $\boldsymbol{J}_{Rp}$ but this time keeping the last three rows of $\boldsymbol{J}_R$ in (4.12). The Jacobian matrix of the uncontrolled state variables is

$$\overline{\boldsymbol{J}}_\phi = 2\boldsymbol{e}_\phi^\top \overline{\boldsymbol{J}_{R\phi}}, \tag{4.65}$$

with $\overline{\boldsymbol{J}_{R\phi}}$ obtained as $\overline{\boldsymbol{J}_{Rp}}$ but using the last three rows of $\boldsymbol{J}_R$ in (4.12).

Remarks similar to the positioning case concerning the number of required DoFs, the singularity of the task Jacobian matrix, and the direct visual measurement of the gripper orientation can be repeated straightforwardly. Notice that these two subtasks are orthogonal.

4.6.3 Onboard-Eye-to-Hand: Camera Field of View

When performing pose-based visual servo (e.g., using the task in Sect. 4.6.2), the target can be easily lost as the task error is computed in the pose space instead of the image plane. Hence, a FoV constraint is essential because the loss of the observed object from the camera image will determine the failure of the whole mission (depending on the available camera optics, this problem could be less significant). We can execute this FoV task together with the positioning task in Sect. 4.6.2 to obtain a hybrid visual servo scheme.

Similar to the image-based visual servo methods presented in Sect. 3.5, let $\boldsymbol{s} = [u\ v]^\top \in \mathbb{R}^2$ be the image features vector of the projection of the observed target centroid. The camera FoV task consists in constraining $\boldsymbol{s}$ within a maximum distance with respect to a desired position $\boldsymbol{s}^d$ in the normalized image plane (e.g., the center of the image) by moving the camera point of view. Without loss of generality, any point of the observed target can be chosen to be controlled in the image. To achieve this goal, we consider the task function defined by

$$\sigma_f = \boldsymbol{e}_f^\top \boldsymbol{e}_f, \tag{4.66}$$

where $\boldsymbol{e}_f = \boldsymbol{s}^d - \boldsymbol{s}$ is the visual servo error as in Sects. (3.5) or (3.6), and the desired task variable is $\sigma_f^d = 0$ (i.e., $\widetilde{\sigma}_f = -\lambda_f \sigma_f$ with λ_f a positive definite gain), while the corresponding task Jacobian is

$$\boldsymbol{J}_f = 2\boldsymbol{e}_f^\top \boldsymbol{J}_S \boldsymbol{J}_{R_f}, \tag{4.67}$$

where $\boldsymbol{J}_S$ is the image Jacobian for the image feature chosen, obtained from (3.26) or (3.42) depending on the image-based visual servo method used, and $\boldsymbol{J}_{R_f}$ is the Jacobian of the robot from (4.8). Both $\boldsymbol{J}_S$ and $\boldsymbol{J}_{R_f}$ are formed only by those columns corresponding to the controllable UAM DoFs.

The Jacobian of the uncontrolled state variables for this task is

$$\overline{\boldsymbol{J}}_f = 2\boldsymbol{e}_f^\top \overline{\boldsymbol{J}}_S \overline{\boldsymbol{J}}_{R_f}, \tag{4.68}$$

with $\overline{\boldsymbol{J}}_S$ and $\overline{\boldsymbol{J}}_{R_f}$ the visual servo and robot Jacobians as in (4.67), respectively, but this time constructed with the columns of the non controllable DoFs.

Notice that only one DoF is required to accomplish this subtask. In fact, we control the distance of the target centroid with respect to the desired optical ray corresponding to $\boldsymbol{s}^d$. However, the corresponding task Jacobian matrix $\boldsymbol{J}_f$ is singular when $\boldsymbol{e}_f \rightarrow \boldsymbol{0}$, but as we do not consider strictly required to move the camera as to have the target object in the precise desired position of the image (i.e., the center), this subtask can be activated only when σ_f exceeds a *safety threshold*. Also a double threshold with hysteresis could be considered in practice to avoid the chattering phenomena when σ_f is close to the threshold. Hence, from a practical point of view the singularity of the task Jacobian is not a problem because this task is deactivated before getting close to a singular configuration.

This subtask is not orthogonal with respect to the gripper positioning but it could be independent if the robot arm is endowed of a sufficient number of joints (i.e., at least 3), with a suitable kinematic configuration.

4.6.4 Eye-in-Hand: End Effector Tracking Using Visual Servo

When the UAM has the camera attached at the multirotor platform, to drive the gripper using visual information we require a hybrid visual servo approach combining both tasks (4.6.2) and (4.6.3). Instead, with an *eye-in-hand* setting, all robot joints intervene in moving the camera. If the arm has more than two DoFs, we can control the camera movements using any of the presented pose-based or image-based visual servoing methods to move the gripper to a desired position and orientation with respect to the object.

Similarly for the other tasks, we can set the task error as the squared norm of the visual servo error, requiring only one DoF to achieve this task (nullifying only the visual servo error norm). However, we prefer to control all camera velocities (i.e., 6 DoFs). Thus, the task error $\boldsymbol{\sigma}_v \in \mathbb{R}^6$ is directly the visual servo error, which depends on the approach used (refer to Sect. 3.3 for all presented visual servo methods). The desired task variable is $\boldsymbol{\sigma}_v^d = 0$ (i.e., $\widetilde{\boldsymbol{\sigma}}_v = -\lambda_v \boldsymbol{\sigma}_v$ with λ_v a positive definite gain), while the corresponding task Jacobian is

$$\boldsymbol{J}_v = \boldsymbol{J}_s \boldsymbol{J}_{R_v}, \tag{4.69}$$

where $\boldsymbol{J}_s$ is the visual servo Jacobian, which depends on the approach (i.e., PBVS, IBVS or UIBVS) and $\boldsymbol{J}_{R_v}$ is the Jacobian of the robot from (4.13). Both $\boldsymbol{J}_s$ and $\boldsymbol{J}_{R_v}$ are formed only by those columns corresponding to the controllable UAM DoFs.

The Jacobian of the uncontrolled state variables for this task is

$$\overline{\boldsymbol{J}}_v = \overline{\boldsymbol{J}}_s \overline{\boldsymbol{J}}_{R_v}, \tag{4.70}$$

with $\overline{\boldsymbol{J}}_s$ and $\overline{\boldsymbol{J}}_{R_v}$ the visual servo and robot Jacobians as in (4.69), respectively, but this time constructed with the columns of the non controllable DoFs.

Notice how in this case, the task function is expressed locally using camera velocities, expressed in the camera frame, obtained from the visual servo approach, and how the task Jacobian converts them to UAM DoFs references.

4.6.5 Center of Gravity

If the arm CoG and the multirotor gravitational vector are not vertically aligned, the motion of the arm produces an undesired torque on the platform base that perturbs the system attitude and position. This effect can be mitigated by minimizing the distance between the arm CoG and the vertical line described by the platform gravitational vector. The task function we introduce is the square norm of this distance, which can be written as

$$\sigma_g = \boldsymbol{d}_{xy}^\top \boldsymbol{d}_{xy}, \tag{4.71}$$

with

$$\boldsymbol{d}_{xy} = \begin{bmatrix} 1 & 0 & 0 \\ 0 & 1 & 0 \end{bmatrix} {}^w\boldsymbol{R}_b \, {}^b\boldsymbol{p}_g, \tag{4.72}$$

where the desired task variable is $\sigma_g^d = \mathbf{0}$ (i.e., $\widetilde{\sigma}_g = -\lambda_g \sigma_g$ with λ_g a suitable positive gain) and ${}^b\boldsymbol{p}_g$ the position of the arm CoG expressed in the body frame b. This position ${}^b\boldsymbol{p}_g$ is a function of the arm joint configuration defined with

$$ {}^b\boldsymbol{p}_g = \frac{1}{\varpi} \sum_{i=1}^{l} \varpi_i \, {}^b\boldsymbol{p}_{g,i} \,, \tag{4.73} $$

where ϖ is the total arm mass

$$ \varpi = \sum_{i=1}^{l} \varpi_i \,, \tag{4.74} $$

and l the number of arm links. ϖ_i is the mass of link i and ${}^b\boldsymbol{p}_{g,i}$ its CoG position, which corresponds to the CoG for the sequence of links i to the end effector, computed with respect to the body frame with

$$ {}^b\boldsymbol{p}_{g,i} = {}^b\boldsymbol{R}_i \frac{\sum_{j=i}^{l} \varpi_j \, {}^b\boldsymbol{p}_{g,j}}{\sum_{j=i}^{l} \varpi_j} \,, \tag{4.75} $$

where ${}^b\boldsymbol{R}_i$ is the rotation of the link i with respect to the body frame. Notice that all these quantities are a function of the current joint configuration $\boldsymbol{\gamma}$.

Then, the differential relationship between the CoG and the arm joint configuration is

$$ {}^b\dot{\boldsymbol{p}}_g = {}^b\boldsymbol{J}_g \dot{\boldsymbol{\gamma}} \,, \tag{4.76} $$

where ${}^b\boldsymbol{J}_g \in \mathbb{R}^{3\times m}$ is the CoG Jacobian, expressed in the quadrotor body frame,

$$ {}^b\boldsymbol{J}_g = \frac{\partial {}^b\boldsymbol{p}_g}{\partial \dot{\boldsymbol{\gamma}}} = \left[{}^b\boldsymbol{J}_{g,1} \cdots {}^b\boldsymbol{J}_{g,m} \right] , \tag{4.77} $$

with ${}^b\boldsymbol{J}_{g,i}$ the individual joint i Jacobian

$$ {}^b\boldsymbol{J}_{g,i} = \frac{1}{\varpi} \sum_{j=i}^{l} \varpi_j \lfloor \boldsymbol{z}_i \rfloor_{\times} {}^b\boldsymbol{p}_{g,i} \,. \tag{4.78} $$

Notice how the resultant linear velocity is scaled by the mass of the partial CoG in (4.78) because the CoG is the average of the multi-mass system with the consequence that high velocities on smaller masses play a smaller role on the total velocity of the CoG.

Finally, the corresponding task Jacobian from the derivative of (4.71), for the controllable DoFs, is defined as

$$\boldsymbol{J}_g = 2\boldsymbol{d}_{xy}^{\top} \begin{bmatrix} 1 & 0 & 0 \\ 0 & 1 & 0 \end{bmatrix} \begin{bmatrix} \boldsymbol{0}_{3\times 4} & {}^{w}\boldsymbol{R}_b \, {}^{b}\boldsymbol{J}_g \end{bmatrix}, \tag{4.79}$$

whereas the Jacobian for the non controllable DoFs is $\overline{\boldsymbol{J}}_g = \boldsymbol{0}_{1\times 2}$.

With this choice, the CoG of the arm is controlled to be aligned with the CoG of the vehicle along the direction of the gravitational force. Notice that we are assuming that the quadrotor is internally balanced. Otherwise, a different equilibrium point should be assigned for the arm CoG.

4.6.6 Desired Arm Configuration

During flight it may be interesting to drive the arm joints towards a desired value γ^d that can be chosen far from an unrealizable configuration and/or close to one characterized by a high manipulability index or suitable with respect to the assigned task. The sum of normalized distances of the position of the i-th joint to its desired configuration is given by

$$\sum_{i=1}^{m} \left(\frac{\gamma_i - \gamma_i^d}{\overline{\gamma}_i - \underline{\gamma}_i} \right)^2, \tag{4.80}$$

with $\overline{\gamma} = \left[\overline{\gamma}_1, \ldots, \overline{\gamma}_m\right]^{\top}$ and $\underline{\gamma} = \left[\underline{\gamma}_1, \ldots, \underline{\gamma}_m\right]^{\top}$ are the high and low joint-limit vectors respectively.

So our task function is selected as the squared distance of the whole arm joint configuration with respect to the desired one

$$\sigma_l = (\gamma - \gamma^d)^{\top} \boldsymbol{\Lambda}_l \, (\gamma - \gamma^d), \tag{4.81}$$

where $\boldsymbol{\Lambda}_l$ is a diagonal matrix whose diagonal elements are equal to the inverse of the squared joint limit ranges

$$\boldsymbol{\Lambda}_l = \operatorname{diag}\left((\overline{\gamma}_1 - \underline{\gamma}_1)^{-2} \ldots (\overline{\gamma}_m - \underline{\gamma}_m)^{-2} \right). \tag{4.82}$$

The desired task variable is $\sigma_l^d = 0$ (i.e., $\widetilde{\sigma}_l = -\sigma_l$), while the corresponding task Jacobian for controllable DoFs is

$$\boldsymbol{J}_l = \left[\boldsymbol{0}_{1\times 4} \; 2 \left(\boldsymbol{\Lambda}_l \, (\gamma - \gamma^d) \right)^{\top} \right]. \tag{4.83}$$

The uncontrolled state variables do not affect the accomplishment of this task, hence their Jacobian is $\overline{\boldsymbol{J}}_l = \boldsymbol{0}_{1\times 2}$.

One common choice of γ^d for the joint limit avoidance is the middle of the joint limit ranges, if this configuration is far from kinematic singularities, with

$$\gamma^d = \underline{\gamma} + \frac{1}{2}(\overline{\gamma} - \underline{\gamma}) \,. \tag{4.84}$$

When using a hierarchical control law, due to higher priority tasks, some joints could reach positions far from its desired values. However, when a joint is approaching a dangerous configuration, the corresponding components of (4.81) and (4.83) can be extracted from this task to form a new isolated task that can be activated on the top of the task stack, using the dynamic change of task priorities presented in Sect. 4.4.3. With this policy, if mechanically viable, the system will reconfigure its internal DoFs to achieve all the remaining subtasks until the dangerous condition will disappear and the original priority will be restored but starting from a different system configuration.

4.6.7 Manipulability

During a manipulation task, a useful objective function is represented by the arm manipulability index presented in [55] and described by

$$\sqrt{\det\left({}^t\boldsymbol{R}_b \boldsymbol{J}_{\text{A}} \, ({}^t\boldsymbol{R}_b \boldsymbol{J}_{\text{A}})^\top\right)} = \sqrt{\prod_{i=1}^{r} \mu_i} \,, \tag{4.85}$$

where $\boldsymbol{\mu}$ are the singular values of the symmetric square matrix ${}^t\boldsymbol{R}_b \boldsymbol{J}_{\text{A}} \, ({}^t\boldsymbol{R}_b \boldsymbol{J}_{\text{A}})^\top$, which has rank r (i.e., number of non zero singular values). This measure is proportional to the volume of the manipulability ellipsoid. Then, the manipulability index can be transformed to

$$\frac{1}{\sqrt{\prod_{i=1}^{r} \mu_i}} \,, \tag{4.86}$$

which in fact, to simplify its computation, can be described as

$$\sigma_m = \frac{1}{\prod_{i=1}^{r} \mu_i} \,, \tag{4.87}$$

which has a different derivative behavior but preserves the same minimum. The desired task variable is $\sigma_m^d = \boldsymbol{0}$ (i.e., $\widetilde{\sigma}_m = -\lambda_m \sigma_m$ with λ_m a suitable positive gain). Notice that if we consider the manipulability for a 6 DoFs end effector workspace, an arm with at least 6 joints is required.

The Jacobian $\boldsymbol{J}_m$ of the controlled stated variables is defined by

$$\boldsymbol{J}_m = \left[\mathbf{0}_{1\times 4}\ \boldsymbol{J}_\mu\right], \tag{4.88}$$

where $\boldsymbol{J}_\mu$ is the Jacobian of (4.87) that can be computed by following the chain rule for derivatives, and its description is here avoided for the sake of simplicity as it involves very large terms. In fact, its analytical computation requires a huge work and it is usually implemented with a numerical solver. The individual Jacobians of the eigenvalues $\boldsymbol{\mu}$ can be obtained as in [46]. As the multirotor platform does not influence the arm manipulability, the Jacobian of the non controllable DoFs is $\overline{\boldsymbol{J}}_m = \mathbf{0}_{1\times 2}$. An alternative formulation of the cost function can be obtained by applying the gradient based method from [58] or the inverse condition number from [54].

4.6.8 Velocity Minimization

We can apply different weights to the joint velocities in order to arbitrarily distribute the motion on the UAM joints. To do so, we can define an individual task for each controllable DoFs (i.e., defining $m + 4$ different tasks), where generically the corresponding objective function is

$$\sigma_v = \dot{e}^2, \tag{4.89}$$

with $\dot{e} = \dot{q}^d - \dot{q}$, where $\dot{q}^d$ is a desired joint velocity which can vary depending on the evaluated DoF. The desired task variable is $\sigma_v^d = \mathbf{0}$ (i.e., $\widetilde{\sigma}_v = -\lambda_v \sigma_v$ with λ_v a suitable positive gain, which can be different for each joint task).

Then, the task Jacobians $\boldsymbol{J}_v$ correspond to a matrix of size $1 \times (m+4)$ where all terms are null except for the column corresponding to the task DoF, that contains the value of $2\,\dot{e}$. In all task definitions the Jacobians of non controllable state variables are null ($\overline{\boldsymbol{J}}_v = \mathbf{0}_{1\times 2}$).

These cost functions are useful to assure the convexity of the problem.

4.6.9 Quadratic Programming Specific Tasks

The quadratic programming approach presented in Sect. 4.5 allows us to specify tasks not only in task velocity space but also with accelerations. If we want to use these tasks with a control law set in the velocity space as the hierarchical control laws presented in Sect. 4.4, we have to integrate them, for example, with an Euler integration. We describe in the following some examples of these type of tasks which can be useful for UAMs.

4.6.9.1 Limiting Quadrotor Accelerations

When rapid end effector motions are required, it is better to distribute the motion in the arm joints instead of the platform. This goal is achieved by penalizing quadrotor accelerations with

$$\sigma_a = {}^w\ddot{\boldsymbol{p}}_b^{\top}\, {}^w\ddot{\boldsymbol{p}}_b \,. \tag{4.90}$$

In fact, quadrotor accelerations in the horizontal plane are obtained through platform tilting, which can affect other tasks if it is not compensated by the inner control loop. In this case, $\boldsymbol{H}_a$ and $\boldsymbol{m}_4$ in (4.47) turns out to have an identity matrix in the 3×3 upper left block and to be null, respectively.

4.6.9.2 Forces on Quadrotor Horizontal Plane

Limiting the forces exerted by the robotic arm on the quadrotor horizontal plane is beneficial because the vehicle cannot oppose them without tilting, due to its under-actuation. Thus, we can consider the following objective function

$$\sigma_h = \frac{1}{2}\ddot{\boldsymbol{\rho}}^{\top}\, {}^b\boldsymbol{J}_g^{\top}\, {}^b\boldsymbol{J}_g \ddot{\boldsymbol{\rho}} - \ddot{\boldsymbol{\rho}}^{\top}\, {}^b\boldsymbol{J}_g^{\top} \sum_{i=1}^{m} \frac{\partial\, {}^b\boldsymbol{J}_g}{\partial q_i} \dot{\boldsymbol{\rho}} \,. \tag{4.91}$$

In this case, we can compute the terms $\boldsymbol{H}_h$ and $\boldsymbol{m}_h$ in (4.47) by direct inspection of (4.91). This cost function penalizes the inertial forces exerted by the arm, by considering the acceleration of its center of mass.

4.7 Validation and Experimental Results

In this Section validations of the presented task control techniques through simulation case studies and real UAM experiments are presented. It is divided in three subsections, corresponding to the techniques shown in Sects. 4.4.1, 4.4.2 and 4.5, respectively.

For all cases, the overall mission consists on three phases: First the UAM is driven autonomously (taking-off and following waypoints) to a point where a main target is in the camera field of view. Secondly, the task control law is switched on to perform the servoing until a certain error in camera pose is reached by the end effector. Finally when the servoing phase is accomplished, the UAM is autonomously commanded to land close to the initial take-off zone. From these phases, our focus is in the middle one, consisting on the vehicle navigation using visual information and where the task control is active.

For the simulation case studies, we take advantage of ROS [48] running on top of the GAZEBO[1] physics engine. The vehicle dynamics is in all cases based on a modification of the Hector quadrotor stack [37] but with specific parameters set according to the real platforms used in the experiments. These UAMs consist on the two robots presented in Fig. 4.1.

4.7.1 HTPC Using Full Least Squares

The validation of this method has been performed with a particular application in mind, consisting on the grasping of a small light-weight carbon fiber bar and its plugging into a structure fixed to the ground. To do so, we take advantage of an *onboard-eye-to-hand* setting with the UAM presented in Fig. 4.1a. In both simulations and real experiments, the UAM weighs 5kg and it is equipped with a downward looking camera at 25Hz and a 6-DoF robot arm plus a gripper attached to the end effector. The camera is located 50cm ahead the vehicle base with an inclination of 30deg with respect to the vertical axis in a way to observe the grasping and plugging maneuvers without self-occlusion. The target object is a bar with two visual markers at the ends, used to obtain the position and orientation of the target object with respect to the camera. To drive the UAM, we employed a velocity control as in [32, 34]. Video 4 in Appendix 5.A reports real experiments with grasping and pluggin maneuvers using this hierarchical control law.

4.7.1.1 Simulations

The proposed approach has been tested in a simulator developed in the ARCAS[2] project based on Gazebo physics engine—see left frame in Fig. 4.3.

To validate this method, the part of the mission where the task control law is switched on is composed of two main sub-phases:

- *Approaching*—the UAM starts from a distance of about 125 cm and has to move the gripper to a pre-grasping pose at 10 cm over the grasping pose;
- *Grasping*—once the intermediate pose has been reached with an error less than a suitable threshold (2 cm for the position and 2° for the orientation), the target pose is moved towards the final grasping pose in 10 s; the closing of the gripper is then commanded when the final pose has been reached with a good accuracy (1 cm for the position and 1° for the orientation).

We defined four task-stack configurations for the simulation of this hierarchical control law:

(i) We use only the end effector tracking task (PBVS) presented in Sect. 4.6.2.

---

[1] http://gazebosim.org.

[2] www.arcas-project.eu

Fig. 4.3 Bonebraker UAM employed during the simulations (left) and real experiments (middle), and images from the on-board camera during the approaching phase (top-right) and at the plugging instant (bottom-right)

(ii) Apart from (i), we add a secondary task to keep the target in the field of view (see Sect. 4.6.3), converting the whole control law to an hybrid visual servo (HVS).
(iii) Together with (i) and (ii), we set the arm end effector alignment task from Sect. 4.6.5.
(iv) In addition to all of the above, we add the task presented in Sect. 4.6.6 to drive the arm to a desired configuration during flight. Specifically, in this case the arm configuration is set to reach the middle position for all joints, following (4.84).

The achieved results are shown in Fig. 4.4 with different colors for the four considered task-stack configurations. A dashed vertical line is employed to highlight the end of the approaching phase and the starting of the grasping phase. Notice that the approaching and the grasping phases in all the considered case studies have different durations depending obviously on the selected control behavior (i.e., the active tasks stack).

Figure 4.4a shows the time history of the position error norm during the task execution for each case study. For all cases, a smooth nullification of the pose error is observed. In particular, during the approaching phase the position error decreases almost linearly due to the saturation of the maximum vehicle cruise velocity (10 cm/s in these case studies). Figure 4.4b shows the time history of the norm of the orientation error. The initial orientation is only a few degrees far from the final grasping pose, hence the error goes under the threshold in few seconds in all the case studies. Notice how the behavior of both the position and orientation errors are similar in all the cases coherently with the hierarchical task combination adopted in the proposed formulation, i.e. the activation of subtasks cannot affect significantly the behavior of the higher priority tasks.

Figure 4.4c shows the results achieved with the activation of the camera FoV subtask. In detail, this subtask is dynamically activated and deactivated by comparing the error squared with a double threshold, i.e. with a suitable hysteresis (20 ± 2 cm) to avoid chattering phenomena. By taking into account the camera pose with respect to $\mathscr{B}$, the desired position of the image features centroid has been chosen equal to

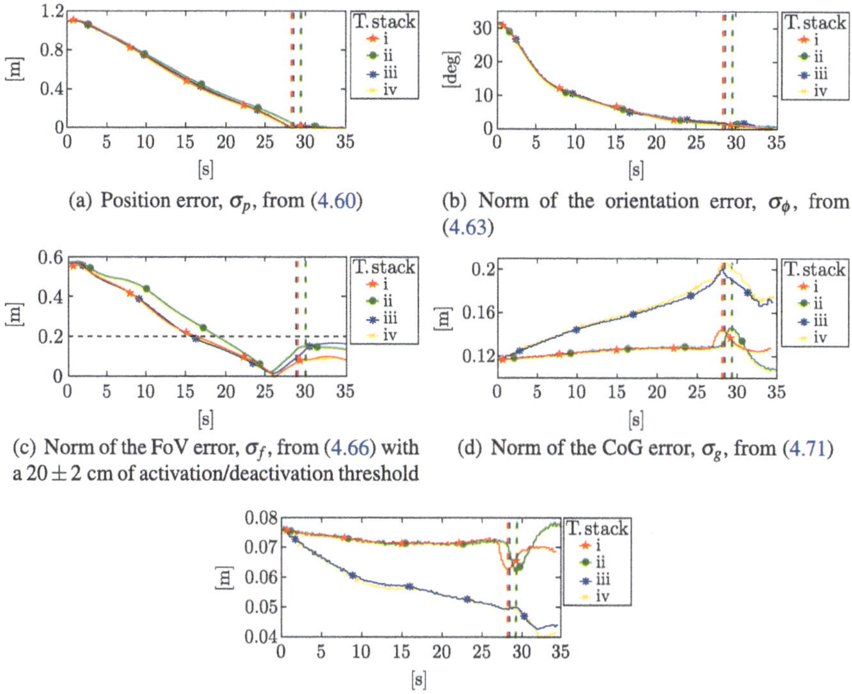

(a) Position error, σ_p, from (4.60)

(b) Norm of the orientation error, σ_ϕ, from (4.63)

(c) Norm of the FoV error, σ_f, from (4.66) with a 20 ± 2 cm of activation/deactivation threshold

(d) Norm of the CoG error, σ_g, from (4.71)

(e) Minimum distance from the joint limits normalized to the joint range, σ_l, from (4.81)

Fig. 4.4 Results of the simulated mission consisting on the approaching and grasping phases. The colored lines in each plot correspond to the task-stacks (i) to (iv). The vertical lines indicate the conclusion of the approaching phase, while the end of each trajectory indicates the grasping time

$s_c^d = \left[0, -0.1\right]^\top$. The achieved results show how except for case (1), i.e. when this subtask is activated, the FoV error is improved without affecting the movement of the gripper.

Figure 4.4e shows the time histories of the error norm for the CoG subtask. For the chosen initial arm configuration the distance of the CoG with respect to the vehicle gravitational axis is 7.6 cm. In cases (1) and (2) this distance remains almost constant, while when the CoG subtask is active, i.e. for cases (3) and (4), the behavior is always improved without affecting the tasks with a higher priority in the stack.

Finally, in the last case study (4) also the joint-limits avoidance constraint is activated. In contrast to the other cases, as shown in Fig. 4.4d the task is not completely fulfilled, even if a clear increase of the distance with respect to the closest joint limit is guaranteed, zero indicates the reaching of a joint limit, while 0.5 indicates that all joints are in the middle of the joint range. This behavior is mainly due to the conflict with other subtasks that have a higher priority in the tasks stack. As described before, it is possible to increase the priority of this task in the stack when a joint limit is excessively close in a way to guarantee mechanical safety at the expense of other tasks.

Table 4.1 Denavit-Hartenberg parameters for the Bonebreaker UAM arm shown in Fig. 4.1a and used in the experiments of Sect. 4.7.1

Link i	γ_i [rad]	d_i [m]	a_{i-1} [m]	α_{i-1} [rad]
1	γ_1	0	0	0
2	γ_2	0	-0.007	$\pi/2$
3	$\gamma_3 + \pi/2$	0	0.2468	0
4	γ_4	0.1105	0	$\pi/2$
5	γ_5	0	0	$-\pi/2$
6	γ_6	0.1113	0	$\pi/2$

4.7.1.2 Real Robot Experiments

The UAM employed for the experimental tests has been developed in the ARCAS[1] project. It is a multirotor aircraft with eight rotors in coaxial configuration with a 105cm tip-to-tip wingspan, height of 50 cm, 13-inches propellers, and a total mass of 8.2 kg including batteries and 6 DoFs robotic arm—see Fig. 4.3. The employed autopilot has been developed by FADA-CATEC[3] and allows also the control of the robot arm. A model-based design methodology established on MATLAB/SIMULINK code generation tools has been adopted [50]. The UAM has been equipped with an i7 ASCTEC MASTERMIND on board for costly computing code, such as image processing. A motion capture system running at 100Hz has been used as the positioning system, while the attitude is measured with the on-board IMU. A 6 DoFs manipulator running at 50 Hz is attached below the vehicle base [8]. The robotic manipulator direct kinematic model is obtained by using the well known Denavit-Hartenberg convention—see Table 4.1.

A high-definition camera running at 14 Hz has been positioned as in the simulation case study. The calibration of the vision system has been divided in two steps. First, the camera intrinsic parameters are obtained with several views of a calibration pattern (i.e., a chessboard). Secondly, the extrinsic parameters are obtained using the motion capture system to precisely localize the platform body frame (b) and an object in the scene (which corresponds to a marker). By knowing the pose of the camera attached to the quadrotor body frame, we can trivially obtain the frame transformation between the camera and the object. However, the estimation of the error between the camera and the optical frames is also required. The marker detector is employed estimating the marker pose with respect to the optical frame. Then, a pose average of the difference between the camera and the optical frames is computed with respect to the object.

Figure 4.5 shows the error between the detected bar and the ground truth poses during grasping and plugging tasks.

The experimental mission consists in plugging a bar with two clipping systems at the ends into a fixed base, as shown in the bottom part of Fig. 4.3. As for the simulated

[3] www.catec.aero.

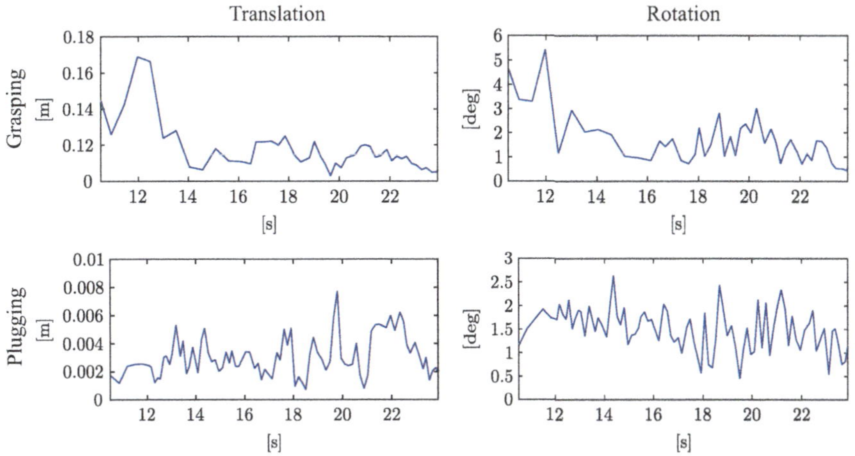

Fig. 4.5 Norm of the object position errors with respect to the ground-truth during grasping and plugging maneuvers

case studies, the mission was decomposed into two steps: the approaching phase, to move the bar over the plugging base at a distance of 5 cm, and the final plugging phase. During this latter phase the FoV task is turned off because the constraint is always satisfied by the system mechanical configuration and the adopted optics. The task requires high accuracy both in position and orientation (i.e., about 1 cm for the position and 1° for the orientation), which has to be guaranteed stable in time to avoid undesired collisions. To cope with this requirement, the bar has been equipped with visual markers as for the plugging base. Hence, the positioning error has been computed by using the measurement of the bar and of the base in a way to mitigate the effects of the calibration errors.

The achieved results are shown in Fig. 4.6. Plots (a) and (b) show the time history of the norm of the position and orientation errors, respectively. The vertical dashed line indicates the end of the approaching phase and the beginning of the plugging phase. The plugging instant corresponds with the end time of the plots. One can observe how the initial errors are quite high because the system starts from a distance of about 40 cm from the goal position, and with a significant orientation error too, however for both errors the target accuracy has been reached in a fast and stable way.

The time history of the FoV error σ_f is shown in Fig. 4.6c, from which one can observe how this subtask is suitably executed, hence the system is able to prevent the loss of the visual markers from the camera FoV.

The CoG subtask has been employed with an activation/deactivation threshold of 15 ± 2 cm. However, it is never activated because the high-priority FoV subtask determines arm configurations already compatible with the CoG subtask. In fact, the alignment error of the CoGs is lower than 4 cm.

Finally, Fig. 4.6e shows the minimum distance computed over all the arm joints between the desired and current positions (zero indicates the reaching of a desired joint position, while 0.5 indicates that all joints are in the middle of the joint range).

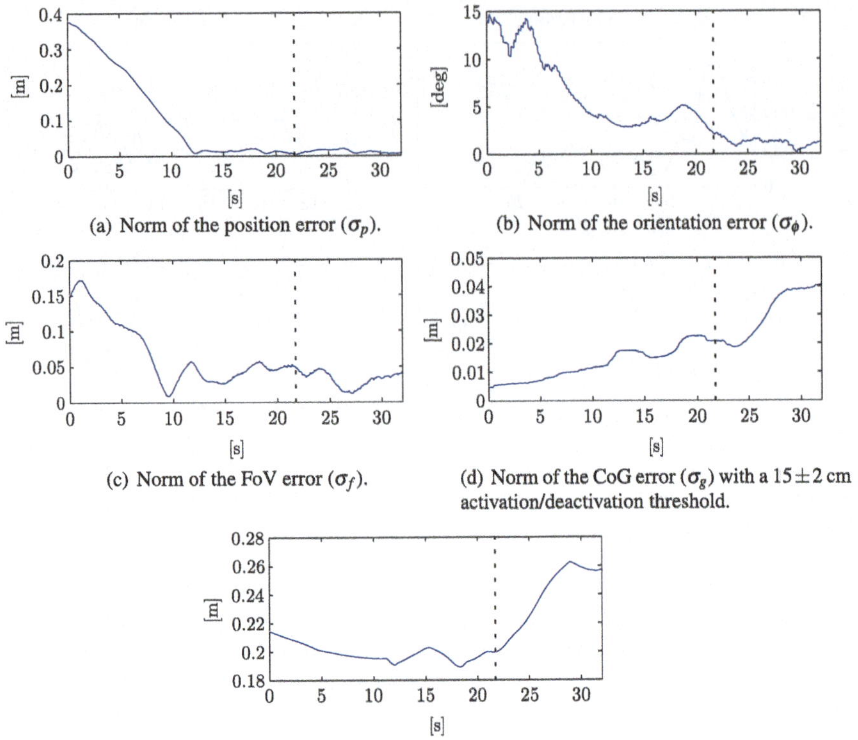

(a) Norm of the position error (σ_p).

(b) Norm of the orientation error (σ_ϕ).

(c) Norm of the FoV error (σ_f).

(d) Norm of the CoG error (σ_g) with a 15 ± 2 cm activation/deactivation threshold.

(e) Minimum distance from the joint limits normalized to the joint range.

Fig. 4.6 Experimental results of the plugging task. The vertical dotted lines indicate the time instant when the approaching phase is concluded, while the each trajectory ends at the plugging time

Even if this is the lower priority task, a safety distance of more than 20% of the joint ranges, in the worst case, is always preserved.

4.7.2 HTPC Decoupling Algorithmic Task Singularities

The validation of this method has been done with a similar mission than in Sect. 4.7.1 but this time, without loss of generality, we have not considered the particular grasping and plugging operations. Here, we take advantage of an *eye-in-hand* setting with the UAM presented in Fig. 4.1b. In both simulations and real experiments, the UAM weighs 1.5 kg and it is endowed with a 6 DoF robot arm plus a camera at 25 Hz attached to its end effector. Simulation case studies and real experiments are descibed in the following and respectively included in Video 5 and 6 in Appendix 5.A.

We designed and built a light-weight robotic arm with a joint setting to compensate the possible noise existing in the quadrotor positioning, while the quadrotor is

Table 4.2 Denavit-Hartenberg parameters for the Kinton UAM arm shown in Fig. 4.1b and used in the experiments of Sect. 4.7.2

Link i	γ_i [rad]	d_i [m]	a_{i-1} [m]	α_{i-1} [rad]
1	γ_1	0	0	0
2	$\gamma_2 - \pi/2$	0	0	$-\pi/2$
3	$\gamma_3 - \pi/2$	0	0	$-\pi/2$
4	γ_4	0	0.065	0
5	$\gamma_5 + \pi/2$	0.065	0.065	0
6	γ_6	0	0	$\pi/2$

hovering in the desired position. The arm was designed to occupy a minimal space in its initial configuration to avoid collisions during take off and landing maneuvers. However its design was a trade-off between accuracy and payload, leading to a weight of 200 g including batteries and approximately 0.2 rad and 10 mm of precision error with the end effector in the worst case (e.g., with the arm at its highest motor torque capacity). The arm is shown in Fig. 4.1b and its Denavit-Hartenberg parameters are given in Table 4.2. The camera is displaced 20 mm from the end effector along the z axis (defining the transform ${}^{t}\boldsymbol{T}_c$).

Now, the target consists on a generic visual marker to easily obtain its position and orientation with respect to the camera, and in the real experiments, we also show a target detected directly using a motion capture system.

In order to maximize the chances for mission fulfillment while guaranteeing robot integrity, in this Section we consider the following task-stack order strategy:

- Robot integrity and safety tasks: stability, collisions, motor torque and velocity limits, etc.
- Mission task. We should ensure enough DoF for the mission in the majority of cases. In case of lacking DoF, the robot integrity prevails.
- Comfort and secondary tasks. It is possible, even probable, that the robot lacks some DoF for these tasks, but at least it tries to accomplish them.

To validate this hierarchical control law, we use the following ordered tasks: a primary safety task (o), presented in Sect. 4.6.1, considering potential collisions (obstacle avoidance); a secondary task, described in Sect. 4.6.4, performing an end effector tracking by using an uncalibrated image-based visual servo (s); and lower in the hierarchy, the alignment of the center of gravity of the UAM (g), shown in Sect. 4.6.5, and a technique to drive the robot arm to desired joint configurations (l) presented in Sect. 4.6.6.

4.7.2.1 Simulations

Similarly to Sect. 4.7.1.1, we present now simulations in ROS [48] and using the Gazebo simulator, but this time the model parameters are according with an Asctec

Pelican quadrotor and those of the robotic arm presented in Table 4.2. The simulated UAM model is shown in Fig. 4.7 (top row).

The visual servoing scheme presented consists on two least squares minimizations. First we solve for the control point coordinates in camera frame albeit scale in (3.30). Then we use the inter-distance constraints to solve for scale and focal length in (3.35). As explained in Sect. 3.6.1, we assume a set of randomly selected 3D feature points on the target and their 2D projections. Instead of developing a robust 3D feature detector and tracker, we us a planar detector of a target of a known geometry to retrieve the target frame to which we add virtual features and then compute the location of these points with respect to the target frame, as well as their basis, i.e., the control points. At each iteration, the marker is detected in the scene and the projection of the control points is computed. This is schematically shown in Fig. 4.8 with the randomly selected feature points (orange dots), the four control points that constitute the basis (violet points) and the respective 2D projections (yellow dots). Those 2D-3D feature relationships represent the input to our visual servoing algorithm (3.30). Notice how our scheme can be applied also to other sensory setups that can detect the object frame from other sources and not just visual features. In the accompanying video we show real experiments using the marker detector shown in Fig. 4.7 (middle row), as well as with an Optitrack system to detect the target frame shown in Fig. 4.7 (bottom row).

Among all other tasks, here we choose the safety task to have the highest priority, not to compromise the platform integrity. Figure 4.9 shows an example of how this task works. We start the servo in a point free of collisions. The inflation radius is set to 0.5 m and at the middle of the expected trajectory we add an obstacle 0.2 m to the left of the quadrotor. Notice how in Fig. 4.9 the safety task becomes active (vertical gray areas in all plots) on the quadrotor y axis to move it away from collision. This DoF is used to keep the platform integrity and cannot be used in other tasks with lower priorities including the main visual servo task. When the obstacle does not violate the inflation radius, the safety task becomes deactivated and the other subtasks can regain access to the previously blocked DoF. Figure 4.9a shows how the servo task is elusive during the first 10 s of the simulation when the obstacle is present, but is accomplished afterwards when the obstacle is no longer an impediment to the secondary task. The activation and deactivation of this task can induce some chattering phenomena. Although this is not explicitly considered in the formulation, one can define a hysteresis scheme for the desired task variable σ_o^d (4.52) or exchange the role of the safety measure to a lower priority when the inflation radius is not violated and move it back to the highest priority when required. In this last case also a smoothing procedure such as the one described in Sect. 4.4.3 must be considered to adequately switch priorities.

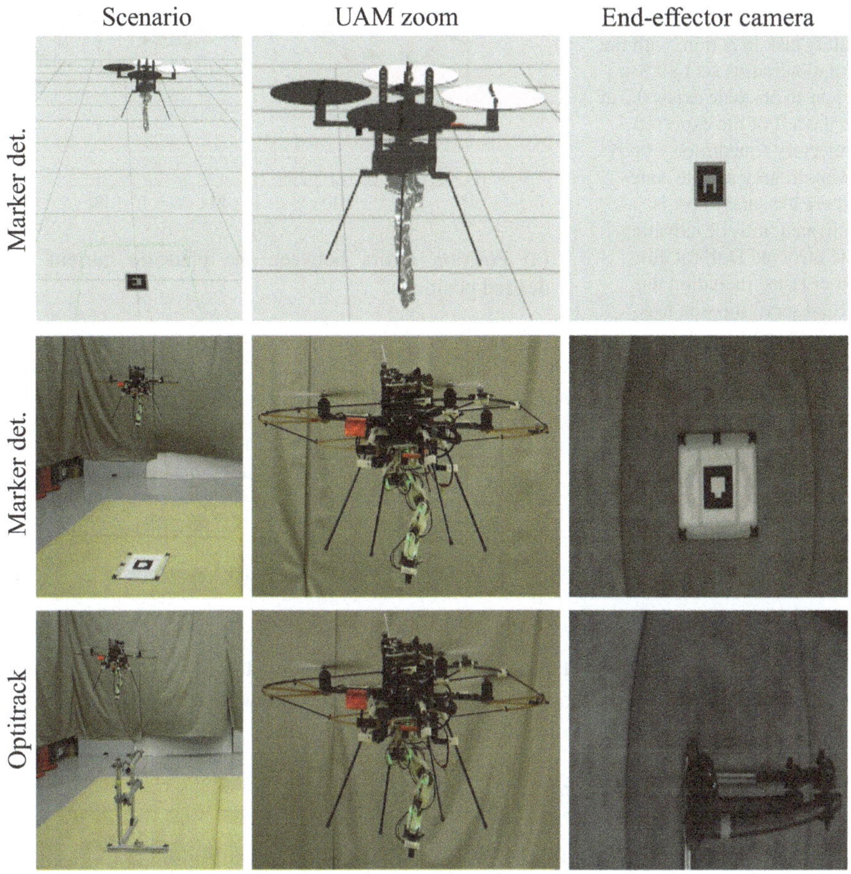

Fig. 4.7 Views during three missions using different object detector techniques: marker detector in simulation (first row), marker detection in a real scenario (second row), and Optitrack tracking (third row)

Fig. 4.8 Object detection scheme with the 3D control points c_j (3.33) in violet, the 3D object features in orange and their 2D projections in yellow

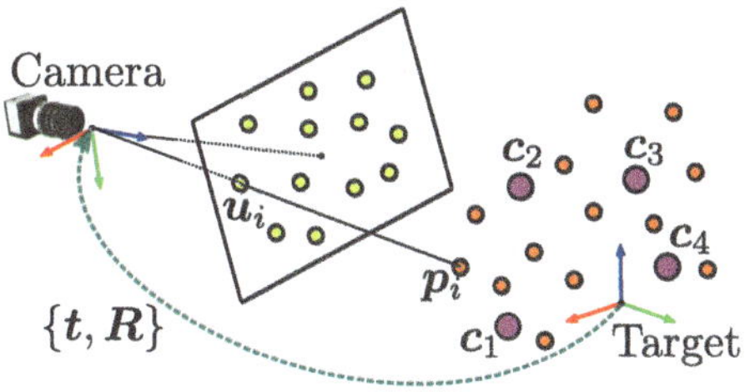

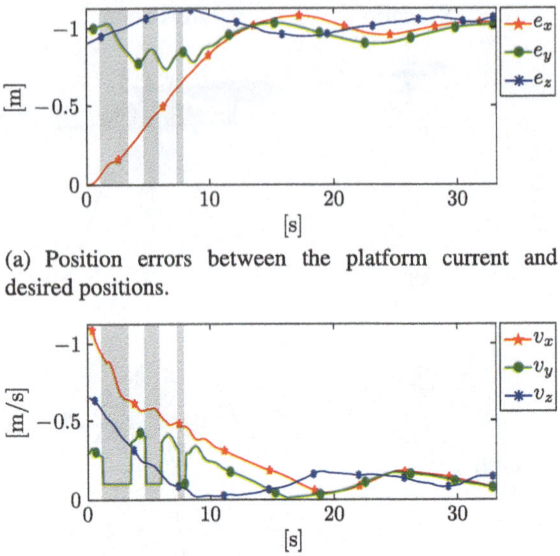

(a) Position errors between the platform current and desired positions.

(b) Visual servo task velocities

Fig. 4.9 Example of the safety task in action, with the inflation radius set to 0.5 m, when an obstacle exists 0.2 m to the left of the expected trajectory (quadrotor y body axis). In grey are the zones where the safety task is activated, thus deactivating the involved DoF for all other tasks, including the visual servo mission task

As with the visual servo task, we can also analyze the stability of the safety task by considering the candidate Lyapunov function of (3.6), whose derivative is given in this case by—see (4.55)

$$\dot{\mathscr{L}} = \boldsymbol{e}^T\, \dot{\boldsymbol{e}} = -\lambda\, \boldsymbol{e}^T\, \boldsymbol{J}_o\, \boldsymbol{J}_o^{\#}\, \boldsymbol{e}\,. \tag{4.92}$$

Then the global asymptotic stability requires the following sufficient condition

$$\boldsymbol{J}_o\, \boldsymbol{J}_o^{\#} > 0. \tag{4.93}$$

By construction, $\boldsymbol{J}_o$ and $\boldsymbol{J}_o^{\#}$ have the same DoF signs and thus, the scalar Ineq. (4.93) holds.

We now compare the effect of using the remaining subtasks in the hierarchy by launching several missions considering the following task stacks:

(i) We use only the 6 DoFs uncalibrated image-based visual servo presented in Sect. 4.6.4.
(ii) Apart from (i), we set the arm end effector alignment task from Sect. 4.6.5.
(iii) In addition to (i) and (ii), we add the task presented in Sect. 4.6.6 to drive the arm to a desired configuration during flight. Specifically, in this case the arm configuration is set to reach the middle position for all joints, following (4.84).

The resulting statistics are presented in Table 4.3 and in Fig. 4.10, which shows the root mean square error (RMSE) between the current and desired camera poses w.r.t. to mission time for the different task compositions.

Table 4.3 Time to completion statistics for multiple realizations of an experiment considering different subtask arrangements

Task stack	Time to target [s]	
	μ_t	$std(t)$
i	42.143	17.361
ii	29.973	12.833
iii	29.036	11.857

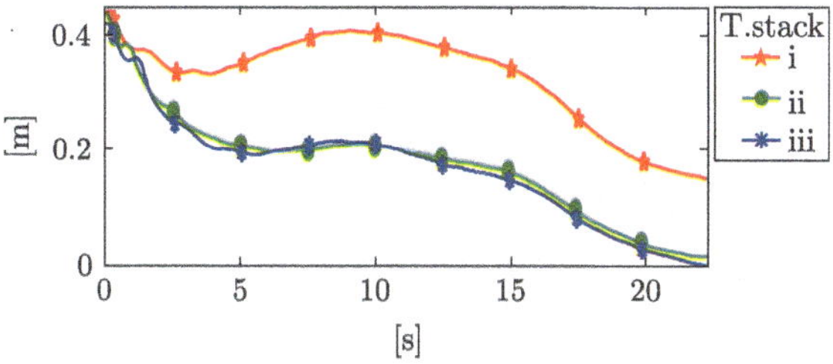

(a) Linear RMSE.

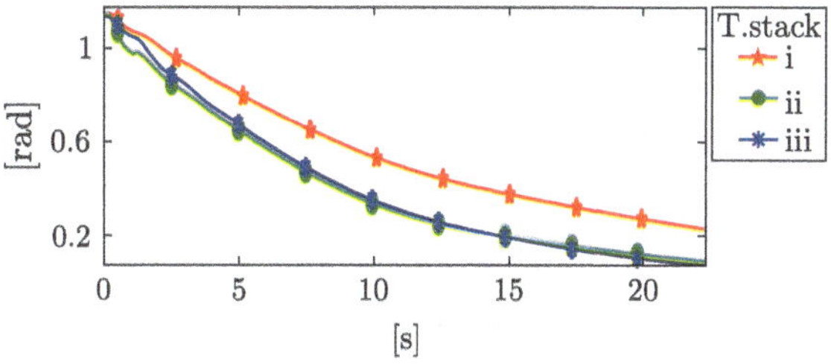

(b) Angular RMSE.

Fig. 4.10 Root mean square error (RMSE) for multiple simulations considering different subtask arrangements

When only the visual servo (i) is executed, the time to reach the target is significantly higher than those cases in which the arm CoG is vertically aligned (ii and iii). This is due to the undesired torque added to the quadrotor when the arm weight distribution is not aligned with the quadrotor CoG. By the addition of the CoG alignment task, this torque is reduced during the servo task. However, if only the CoG is aligned, the arm can still reach undesired configurations, close to singularities or joint limits. The slight improvement in RMSE between (ii) and (iii) are because the arm is fully extended in the (ii) case, increasing the vertical distance between the arm CoG and the platform base, leading to larger inertial effects than the (iii) where a retracted configuration was set as the last task in the hierarchy.

The control law proposed in [28] contains a unique secondary task corresponding to a weighted sum of subtasks, which can be problematic when the tasks are antagonistic. Depending on the weights assigned, the resulting velocities could not satisfy accurately some of the subtask requirements, e.g. when one task tries to reach a singular arm configuration with the arm CoG vertically aligned with the quadrotor gravitational vector, whilst another task is driving the robot away from such singu-

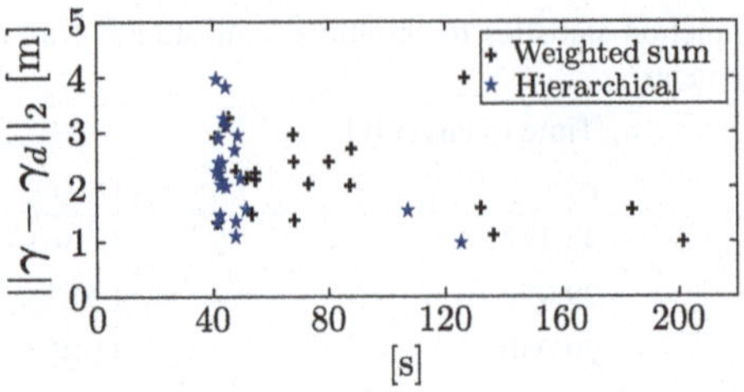

Fig. 4.11 Comparison between weighted sum and hierarchical task composition, considering the time to reach the target (horizontal axis), with final linear and angular Euclidean distances lower than 5 cm and 0.026 rad respectively, and for randomly varying initial conditions (vertical axis)

Table 4.4 Time to completion statistics for multiple realizations of the simulation under varying initial conditions for the two methods: weighted sum, and hierarchical task composition

Method	Time to target [s]	
	μ_t	$std(t)$
Weighted sum	85.3979	46.5853
Hierarchical	50.9624	21.4597

larity and hence, from CoG alignment. The resultant velocity from the sum of the two would drive the camera to a pose that does not satisfy either of the two tasks and still the weighted sum would be reduced to zero. In contrast, the control law presented here takes into account the priority of each task. That is, the desired arm configuration of the last task will be only fulfilled if it does not bring the arm and quadrotor CoGs away from vertical alignment.

To show the advantage of hierarchical task composition against the weighted sum method we performed extensive simulations with the two strategies for varying initial conditions and final desired configurations. The results are shown in Fig. 4.11 and in Table 4.4. In all cases, simulations were ceased once a distance to the target smaller than 5 cm was reached with an orientation closer than 0.026 rad. The main observed result is that both strategies were equally capable of reaching the target with the desired accuracy level, and that the hierarchical task composition method consistently did so in about 50 s, independent of the initial configuration; whereas the weighted sum method required on average 85 s to achieve the task. We can conclude that the proposed method reaches task completion sooner than the method presented in [28] mainly because it prioritizes quadrotor stability through CoG alignment over the last task, thus avoiding antagonistic task behaviors.

4.7.2.2 Real Robot Experiments

We conducted a series of experiments with missions similar to those shown in the simulations, i.e., autonomously taking off and flying to a location in which the target appears in the field of view of the camera, turning then on the hierarchical task controller to servo the system towards a desired camera pose, and finally autonomously landing the system. Even when our simulator runs a complete and robust physics engine, it does not account for the most elaborated physical phenomena such as frictions, damping or inertias, raising the need for a real demonstration. One significative difference between our simulated setup and the real setting was that in simulations, the robot was able to finish the servo task even with the rest of the tasks in the hierarchy inactive—see Table 4.3. In the real case however, task completion was elusive without making use of the entire task hierarchy.

The real experiments were conducted with our robot Kinton (Fig. 4.1b), based on an Asctec Pelican quadrotor, and equipped with an on-board embedded PC (1.6 GHz CPU) and a variety of sensors including an IMU and a barometer. All our algorithms are running on board in real time with a camera frame rate at 20 Hz.

Figure 4.12 shows a comparison of task execution with and without activation of the task hierarchy. The top frames show linear and angular pose errors when only the servo task is active. Interestingly enough one can observe how in this case the arm wrist rotation is compensated with the quadrotor yaw. These movements are antagonistic and their subtraction cannot be appreciated at the image level. If on the other hand more tasks are added in the hierarchy, such as CoG alignment or joint limits, the hierarchical control law can make use of the floating DoFs to achieve the extra tasks, as shown in the bottom frames in the figure.

In this experiment, task completion is considered to be reached at an Euclidean position error of 0.15 m and 0.2 rad; otherwise task abortion is executed if after 3 min of flight time the target is not reached. We are aware that better performance is possible with a more elaborate tuning of the different control gains, however, this level of precision is good enough to show that the task composition scheme allowed the system to quickly reach its target, whereas without the hierarchical task composition in place, the task could not be accomplished.

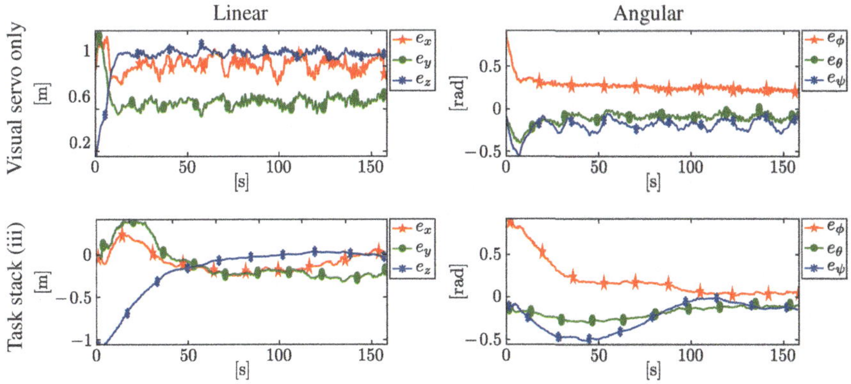

Fig. 4.12 Camera pose error during 6 DoFs visual servoing. Comparison of using or not the hierarchical task priority control law with all the subtasks proposed (i.e., task stack iii)

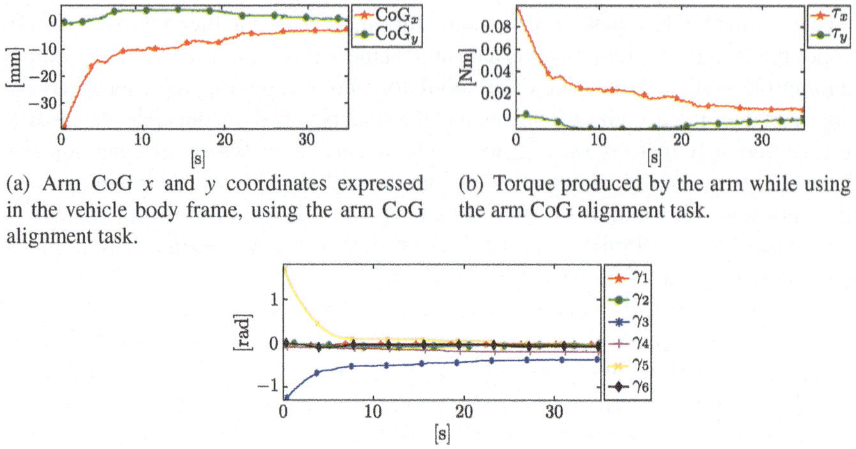

(a) Arm CoG x and y coordinates expressed in the vehicle body frame, using the arm CoG alignment task.

(b) Torque produced by the arm while using the arm CoG alignment task.

(c) Arm joints error with respect to their desired position using the desired arm configuration task.

Fig. 4.13 Effects in the quadrotor body frame of applying the arm CoG vertical alignment task

Arm CoG alignment is crucial to improve flight behavior. With it the approaching maneuver is softer allowing us not only to easily reach the desired servo error but also reducing aggressive maneuvers to compensate the arm payload, thus reducing energy consumption which is a very important issue for this type of aerial platforms. Figure 4.13 shows the effect of this alignment with a comparison of CoG alignment versus overall arm torque.

The last task is designed to favor a desired arm configuration and it can be used to push the joints away from singularities and potentially increase maneuverability. Figure 4.13c shows the error between the current and desired joint positions when the task is included in the hierarchy at the lowest priority level.

Finally, to evaluate the contribution of each control variable to the execution of the different tasks we present plots of the whole set of velocity profiles applied to the UAM actuators (i.e., 3 quadrotor linear velocities and 6 arm angular velocities plus quadrotor yaw velocity) in Fig. 4.14. Note how for the main mission task, the visual servo task, the quadrotor linear velocities play an important role during the first 5 s of the experiment, when the UAM is far from the target, with the arm joints accommodating later on for the fine positioning of the camera thanks to the time-varying weighted motion distribution presented in Sect. 4.3.1.

The fact that all tasks report velocity values for the control variables indicate the availability of DoFs for their execution from their higher priority tasks. Indeed, the dimension of the associated space to each null space projector (i.e., the number of singular values different from zero, which in the case of orthogonal projection matrices are always equal to 1) are $dim(N_o) = 10$, $dim(N_{o|v}) = 4$ and $dim(N_{o|v|g}) = 3$. These dimensions indicate how tasks with lower priority than the inflation radius

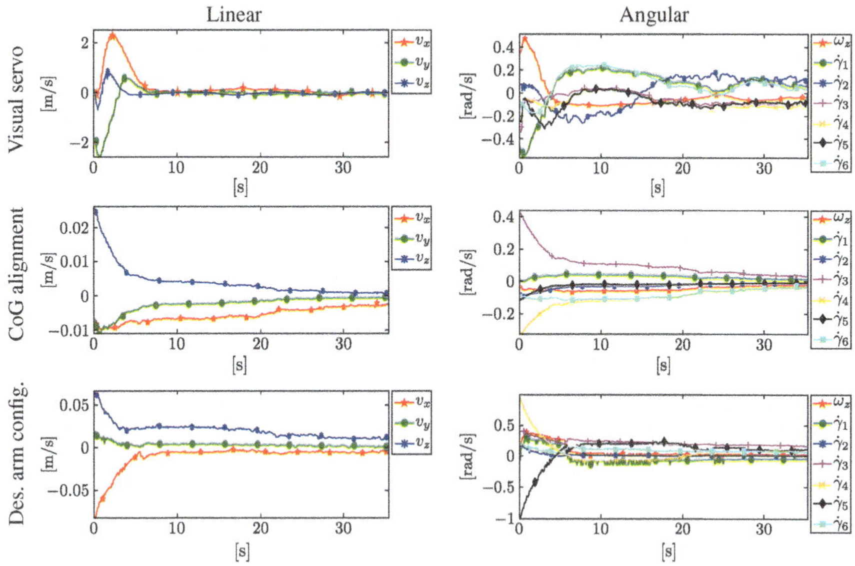

Fig. 4.14 Actuator velocities applied in a real experiment corresponding to the individual contributions of each subtask: 6 DoFs visual servoing, CoG alignment and desired arm configuration

task can actuate on all 10 DoFs of the robot when the inflation radius is not violated. The visual servo mission task requires 6 DoFs, and the secondary and comfort tasks with lower priority can take advantage of the remaining 4 DoFs. The gravitational vector alignment task and the joint limits avoidance task require 1 DoF each being scalar cost functions to minimize—see (4.71) and (4.81). These results have been experimentally confirmed, computing the number of singular values associated to each of the null space projectors in the task hierarchy.

Although the dimension of these associated spaces give an idea of the available DoFs for each task in the hierarchy, it does not imply that the subtask can always be fulfilled. For instance, a subtask requiring a specific non-available joint motion might not be possible even when other decoupled DoFs are still free. An empirical study of this consideration led us to the task order priority presented in the beginning of this Sect. 4.7.2, and a thorough analytical study of these spaces by means of their basis (i.e., singular vectors associated to the non-zero singular values) and their implications to guarantee subtask completion is left as future work.

4.7.3 Optimization-Based Trajectory Generation

The proposed trajectory generation algorithm is implemented on a real UAM, and the effectiveness of the approach is demonstrated by performing an autonomous mission, by accomplishing different tasks while enforcing system constraints.

In this Section only results from real experiments are presented. Although simulations using Matlab, Gazebo and ROS have been performed, their results are here avoided for the sake of conciseness. Notice that we do not compare our performance against other methods because, to the authors knowledge, this is the first work using an optimization based approach to achieve such trajectory generation for UAMs with all algorithms running on board in real time. Video 7 referenced in Appendix 5.A shows Gazebo simulations together with real robot experiments using the QP approach.

The quadrotor used in the experiments is, as in the previous Sect. 4.7.2.2, the UAM presented in Fig. 4.1b. This UAM is composed by an ASCTEC Pelican research platform and the overall control architecture is shown in Fig. 4.15. This platform has a position controller in cascade with an off-the-shelf built-in autopilot for attitude estimation and control, which are not the focus of this paper. As for the robotic arm, its Denavit-Hartenberg has been already presented in Table 4.2. All algorithms are running on board in real time using an Intel Atom CPU (@1.6GHz) with Ubuntu 14.04LTS and ROS Indigo. The optimized high-rate C++ implementation takes advantage of QPoases as quadratic programming solver [13] and is available upon request. All experiments have been performed at Institut de Robòtica i Informàtica Industrial (IRI), CSIC-UPC, equipped with an Optitrack motion capture system running at 120 Hz and used in the low-level quadrotor position control. Figure 4.16 show the experiment setup with the flying arena and the cylindrical pilar used as obstacle. In all the experiments the quadrotor is autonomously taken off and landed, and both maneuvers are considered out of the paper scope.

A subset of cost functions and constraints presented in Sects. 4.6 has been implemented in the real platform. In particular, the main task of the mission consists in the end effector trajectory tracking from Sect. 4.6.2 (only considering a global end effector tracking without camera particularities), while complementary tasks used to solve the system redundancy are the alignment of arm CoG Sect. 4.6.5, the positioning of arm joints to a favorable configuration Sect. 4.6.6, and the minimization of joints velocity Sect. 4.6.8. A constraint avoiding self-collision between robot end

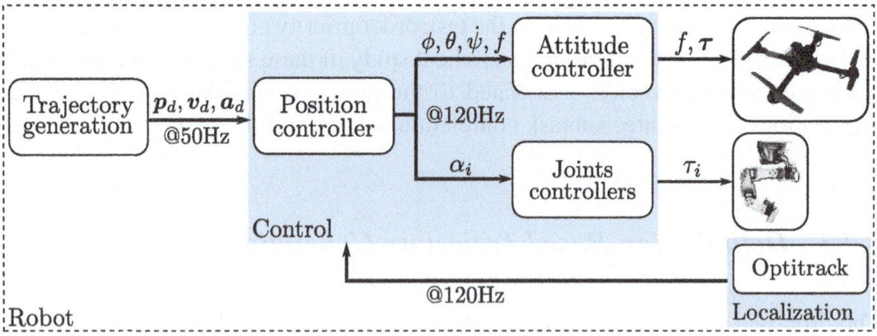

Fig. 4.15 Overview of the architecture pipeline for trajectory generation and UAM control with all algorithms running on board

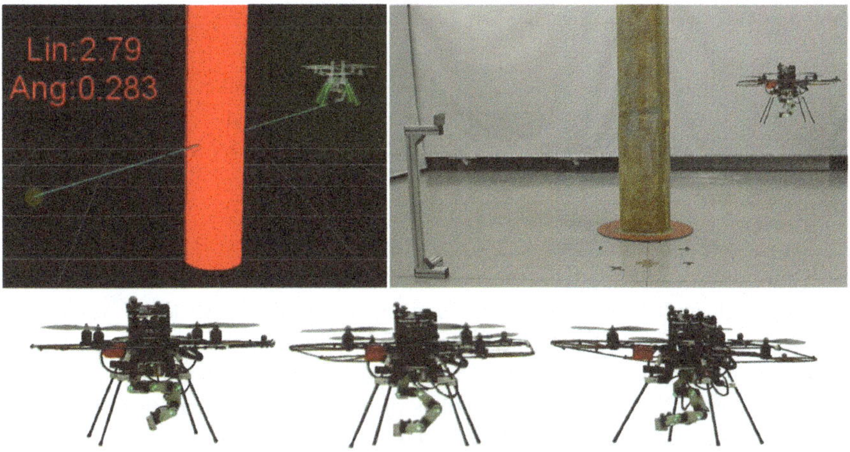

(a) After taking off. (b) During navigation. (c) Close interaction

Fig. 4.16 Experiment setup with the flying arena and a cylinder pilar used as obstacle (the left frame corresponds to the telemetry visualization), to test the task controller using the QP approach. The bottom frames are samples of arm configurations depending on tasks weightings for different mission phases

effector and quadrotor legs is active for all the duration of the experiment Sect. 4.6.1. In addition, the obstacle avoidance constraint is also tested. In particular, an experiment without any obstacle in the path and another one with a cylindrical shape obstacle will be compared. Finally, position, velocity and acceleration of all DoFs are subject to user selected bounds Sect. 4.5.3.

Therefore, we define a desired waypoint for the end effector positioning task (3D position plus 3D orientation), which will drive the whole robot. The waypoint presents a displacement of 2 m in both x and y directions, and 0.2 m in z direction. The mission is considered achieved when the linear and angular positioning errors of the end effector are below certain thresholds. These thresholds are selected considering the hardware characteristics previously mentioned in Sect. 4.7.2, which are 0.05 m and 0.2 rad of linear and angular errors.

In order to show the effects of the different tasks and constraints, we split the part of the mission where the trajectory generation algorithm is active in two sub-phases: navigation and interaction. In the first phase, the navigation towards the waypoint has to be preferably performed with quadrotor DoFs, while arm joints should assume a configuration in order to maximize stability and minimize disturbances, e.g., torques produced by displacement of the arm CoG. In the interaction phase, when the robot is close to the desired waypoint (i.e., almost hovering for close manipulation), it is preferable to perform the motion with arm joints, because more accurate movements are needed and a minimum safety distance with the interaction object can be kept.

The easiness of the proposed method allows to distinguish the different phases by dynamically changing the weights of the different tasks. The proposed normalization

Table 4.5 Tasks weightings depending on mission phases

	Tasks				
	EE	CoG	Des. arm conf.	MinVel	
				Quad.	Arm
Navigation	1	10^1	10^2	10^{-5}	10^{-5}
Interaction	1	10^{-2}	10^{-2}	10^3	10^{-5}

procedure allows to effectively assign a relative priority to the tasks by means of weights, even with tasks of non homogeneous dimensions. All these weights are summarized in Table 4.5. In the subsequent figures, the transition between navigation and close interaction phases is shown with a black vertical dashed line.

As presented in Sect. 4.5, the dynamics of the joints, tasks and the approach to constraints is governed by parameters λ_1 and λ_2. For the described system, they have been chosen equal to 0.8 $1/s$ and 4 $1/s$, respectively, in accordance with lower-level control loop bandwidth.

In Fig. 4.17 the behavior of all UAM joints during the complete experiment with the obstacle is reported. Figure 4.17, frames (c) to (f), shows the task errors during the mission, whereas in Fig. 4.18 the quadrotor trajectories (x and y plot) are shown with and without an obstacle lying in the middle of the shortest path. The different motion profiles are discussed in the following.

4.7.3.1 Navigation Phase

During the navigation, large weights are assigned to the cost functions of CoG alignment and desired joint positions. Thus, long displacements are performed by the quadrotor and the arm is driven to a desired position while minimizing CoG misalignment. These profiles are reported in Fig. 4.17. The behavior of the end effector task error is reported in Fig. 4.17c and d, for the translational and rotational parts, respectively. In Fig. 4.17e and f, the profiles of the CoG alignment and the arm joints positioning can be analyzed. Notice that the CoG task is not completely accomplished, i.e., the task error is not reduced to zero, because the latter has larger weight, and the equilibrium solution is a weighted average between the two goals.

The arm configurations are shown in the bottom frames of Fig. 4.16, where frame (a) is just after the take off, and frame (b) represents the robot configuration during navigation where the CoG alignment and arm joints positioning tasks prevail. Frame (c) represents the arm configuration during the interaction phase, that will be described in the following.

To show the need for the collision avoidance constraint we added an obstacle in the middle of the trajectory (i.e., shortest path). The QP solver generates a feasible trajectory, and the UAM avoids the obstacle with the minimum assigned distance of 0.6 m (inflation radius around the platform). To clearly see how the obstacle avoidance works, we show in Fig. 4.18 a comparison between two trajectories with

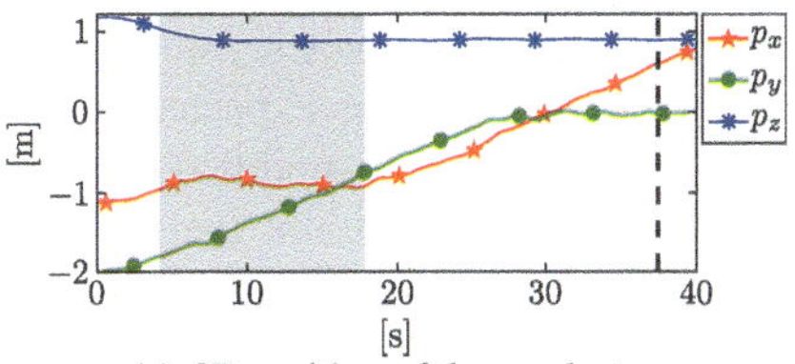

(a) 3D position of the quadrotor.

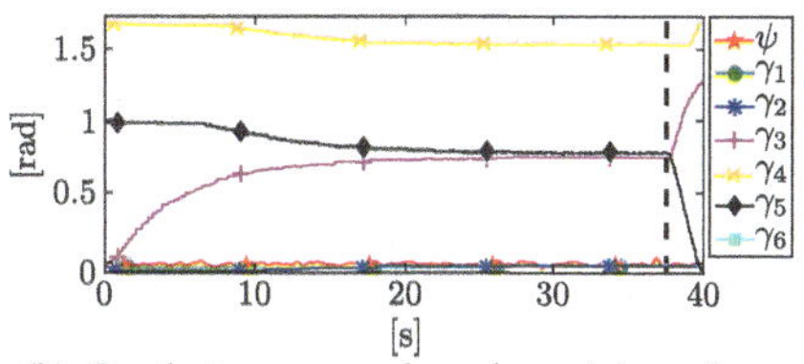

(b) Quadrotor yaw angle and arm joint values.

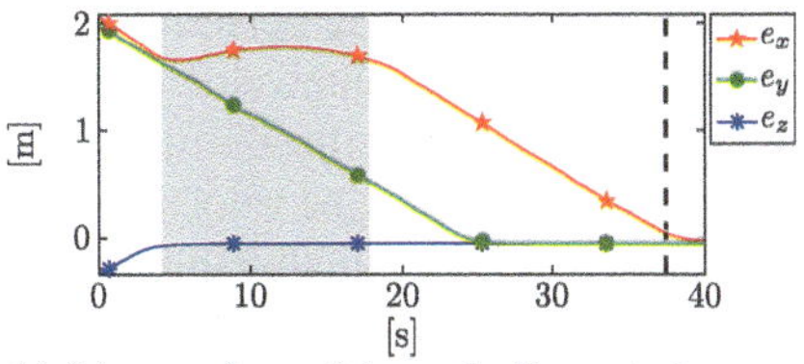

(c) Linear values of the end effector task error, σ_p, from (4.60).

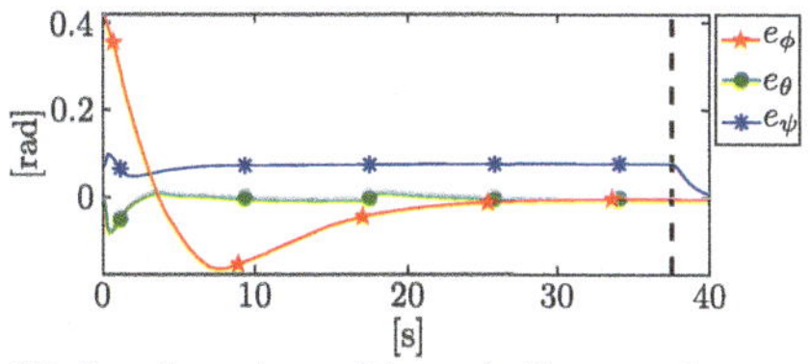

(d) Angular values of the end effector task error, σ_ϕ, from (4.63).

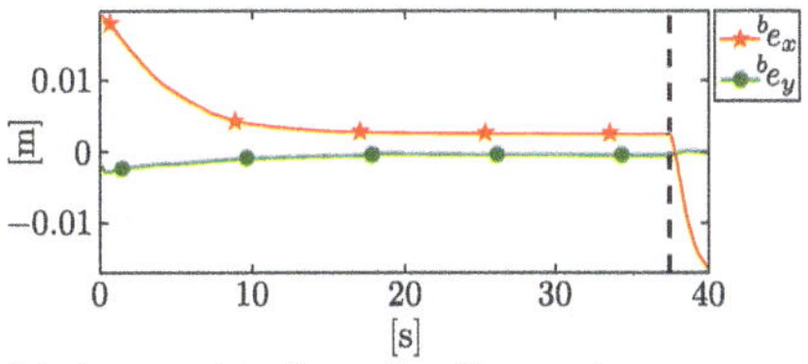

(e) Arm center of gravity alignment error, σ_g, from (4.71).

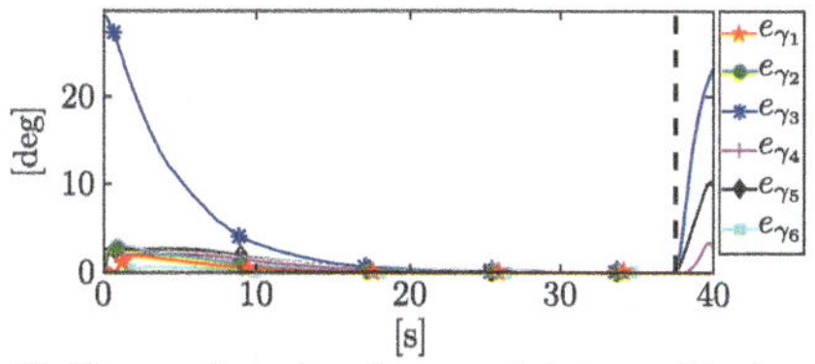

(f) Error values for the arm joints positioning task, σ_l, from (4.81).

Fig. 4.17 Analisis of a real experiment using the QP approach. In this case, there exist an obstacle inline between the initial and desired end effector positions. The gray regions in Fig. (**a**) and (**c**) corresponds to the activation of the obstacle avoidance constraint. Notice how the x axis is blocked (continuous red line with star markers). The vertical dashed line indicates the point where the weights of the arm joints positioning and arm CoG alignment tasks are reduced, and quadrotor motion is penalized

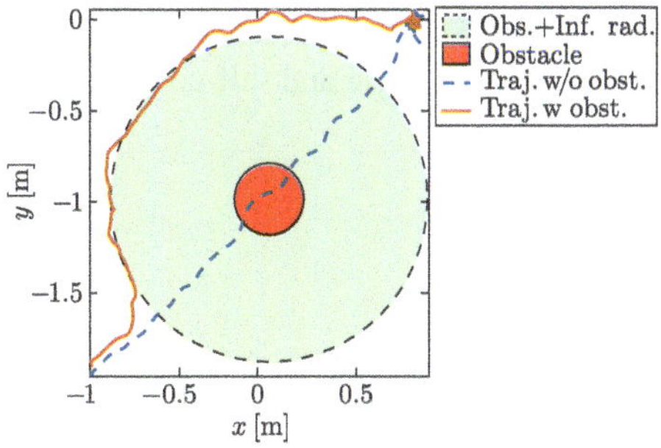

Fig. 4.18 Comparison between two real trajectories with the same initial and final positions, but without any obstacle (dashed blue line) and with an obstacle lying between the initial and desired end effector positions (continuous brown line). The small red circle corresponds to the actual obstacle and the yellow area (yellow circle with a dashed edge) includes the inflation radius applied to the obstacle

the same parameters. A first trajectory is executed without any obstacle (blue dashed line) and the computed path corresponds to the shortest path as expected. When an obstacle appears (defined by the red circle) the trajectory is modified to avoid it (brown continuous line), satisfying the inflation radius constraint (yellow area plotted, in this case, around the obstacle). This behavior is evident in the gray area of Fig. 4.17a, where the motion of the platform is prevented for the x axis (continuous blue line). Once the obstacle has been avoided, the trajectory is resumed.

As it can be seen in Fig. 4.17, the quadrotor motion is clearly performed with constant velocity. In fact, for large end effector errors, the maximum velocity bound is saturated and the quadrotor moves with constant velocity. Notice that the two curves of x and y quadrotor positions have the same slope, because the same velocity bound has been assigned. On the other hand, the z coordinate presents a smaller initial error, thus the velocity is not saturated and, as a result, the behavior is exponential because its dynamics is governed by parameter λ_1. Notice that the z position reference is reached after 6 s and 7 s, while the theoretical settling time $5/\lambda_1$ is equal to 6.25 s.

4.7.3.2 Interaction Phase

The second phase starts when the end effector is 15 cm far from the desired position. At this point, weights are changed to the values of Table 4.5. Notice how the arm joints move towards the goal, see Fig. 4.16c that represents the robot configuration during the interaction phase, and Fig. 4.17, which reports the behavior of the joint variables. As a consequence, CoG and joint positioning task errors are increasing, as reported in Fig. 4.17e and f. The end effector task is performed to a greater extent by robot arm joints, because the weights of the cost functions corresponding to the joint velocity minimization are assigned such that quadrotor motions are more heavily penalized than arm ones. Notice that the trajectory is computed using the acceleration as a regressor and applying bounds to it. For this reason, the transition between the two phases is smooth, without discontinuities.

As a conclusive remark, our UAM can effectively accomplish the mission in the two distinct phases, avoiding an obstacle and self-collisions and respecting variable bounds.

4.8 Conclusions

In this chapter we described several task control architectures suited for UAMs, which take advantage of the high redundancy of these type of aerial robots. Specifically, we presented two hierarchical control laws, together with a trajectory generation algorithm using quadratic programming. Moreover, we defined several tasks specifically designed for UAMs to accomplish during the navigation and close interaction phases in manipulation missions. These tasks were designed with the following purposes: to safeguard the platform integrity by avoiding collisions; to track trajectories with the arm end effector using visual information, with both *onboard-eye-to-hand* and *eye-in-hand* configurations; to keep the target in the camera field of view; to align the arm CoG with the platform gravitational vector; to drive the arm to a desired configuration; to increase the arm manipulability; to force minimum joint velocities; to limit quadrotor accelerations; and to reduce forces on the horizontal plane.

The first hierarchical approach presented, uses a full least squares solution and has been partially published in [29]. The underactuation of the aerial platform has been explicitly taken into account in a general formulation that also guarantees the decoupling of independent tasks. In this case we have showed experiments using hybrid visual servoing to reduce the error of the desired gripper position and orientation. The subtasks in the hierarchy in this case include maintaining the target in the camera FoV; vertically aligning the arm CoG with the quadrotor gravitational vector; and avoiding arm joint limits. Simulations and experiments validated the proposed solution.

A second hierarchical control law was described, this time decoupling algorithmic task singularities. This work has been partially published in [51, 52]. In this case, we presented results using a primary task to avoid obstacles; a secondary task for the uncalibrated visual servo (end effector tracking using an *eye-in-hand* configuration); and lower priority tasks designed to alleviate quadrotor stabilization issues (arm CoG alignment and desired arm joint positions). This hierarchical strategy might not achieve the optimum constraint-task reconstruction errors as in the first control law presented, but instead the algorithmic singularities arising from conflicting tasks are decoupled from the singularities of the secondary tasks. Moreover, the presented control law only requires independent tasks for the uncontrollable variables to guarantee exponential stability of the system.

The presented trajectory generation algorithm is based on a quadratic programming approach and was published in [49]. The method uses an on-line active set strategy to compute feasible joint acceleration references, in order to accomplish given tasks, while enforcing constraints and variable bounds. A number of tasks and constraints specific to unmanned aerial manipulators have been integrated in the algorithm. A weighting factor, associated with a normalization procedure, allows to define the relative importance of tasks, thus exploiting the redundancy of the system. This is important to effectively perform distinct phases of a mission. In particular, the objective functions implemented in this case within the real setup include end effector positioning, the alignment of the arm CoG with the platform gravitational vector, and the positioning of arm joints to a favorable configuration. In addition, we demonstrated that the robot can avoid collisions with known obstacles in the scene and self-collision between end effector and quadrotor legs. To the authors knowledge, this was the first work using an optimization-based approach to compute trajectories for aerial robots with a large number of DOFs, working on board and in real time.

All described tasks and control laws are demonstrated using both simulations and a real UAMs.

References

1. Siciliano, Bruno, and Oussama Khatib (eds.). 2008. *Springer Handbook of Robotics*. Berlin Heidelberg: Springer.
2. Antonelli, G. 2009. Stability analysis for prioritized closed-loop inverse kinematic algorithms for redundant robotic systems. *IEEE Transactions on Robotics* 25 (5): 985–994.
3. Baerlocher, P., and R. Boulic. 1998. Task-priority formulations for the kinematic control of highly redundant articulated structures. *IEEE/RSJ International Conference on Intelligent Robots and Systems*, vol. 1, 323–329. Victoria, Canada.
4. Bellicoso, Carmine Dario, Luca Rosario Buonocore, Vincenzo Lippiello, and Bruno Siciliano. 2015. Design, modeling and control of a 5-DoF light-weight robot arm for aerial manipulation. In *23th Mediterranean Conference on Control and Automation*, 853–858. Torremolinos, Spain, June 2015.
5. Bellman, Richard Ernest, and Kenneth L Cooke. 1963. *Differential-Difference Equations*. Rand Corporation.
6. Bernard, Markus, Konstantin Kondak, Ivan Maza, and Anibal Ollero. 2011. Autonomous transportation and deployment with aerial robots for search and rescue missions. *Journal of Field Robotics* 28 (6): 914–931.
7. Buonocore, Luca Rosario, Jonathan Cacace, and Vincenzo Lippiello. 2015. Hybrid visual servoing for aerial grasping with hierarchical task-priority control. In *23th Mediterranean Conference on Control and Automation*, 617–623. Torremolinos, Spain, June 2015.
8. Cano, R., C. Pérez, F. Pruano, A. Ollero, and G. Heredia. 2013. Mechanical design of a 6-DOF aerial manipulator for assembling bar structures using UAVs. In *2nd RED-UAS, 2013. Workshop on Research, Education and Development of Unmanned Aerial Systems*. Compiegne, France, November 2013.
9. Chan, Tan Fung, and R.V. Dubey. 1995. A weighted least-norm solution based scheme for avoiding joint limits for redundant joint manipulators. *IEEE Transactions on Robotics and Automation* 11 (2): 286–292.
10. Chaumette, François, and Seth Hutchinson. 2007. Visual servo control. II. Advanced approaches. *IEEE Robotics & Automation Magazine* 14 (1): 109–118.
11. Chiaverini, S. 1997. Singularity-robust task-priority redundancy resolution for real-time kinematic control of robot manipulators. *IEEE Transactions on Robotics and Automation* 13 (3): 398–410.
12. Escande, Adrien, Nicolas Mansard, and Pierre-Brice Wieber. 2014. Hierarchical quadratic programming: Fast online humanoid-robot motion generation. *The International Journal of Robotics Research* 33 (7): 1006–1028.
13. Ferreau, H.J., H.G. Bock, and M. Diehl. 2008. An online active set strategy to overcome the limitations of explicit MPC. *International Journal of Robust and Nonlinear Control* 18 (8): 816–830.
14. Forte, Francesco, Roberto Naldi, Alessandro Macchelli, and Lorenzo Marconi. 2014. On the control of an aerial manipulator interacting with the environment. In *IEEE International Conference on Robotics and Automation*, 4487–4492. Hong Kong, China, June 2014.
15. Garimella, Gowtham, and Marin Kobilarov. 2015. Towards model-predictive control for aerial pick-and-place. In *IEEE International Conference on Robotics and Automation*, 4692–4697. Seattle, USA, May 2015.
16. Geisert, Mathieu, and Nicolas Mansard. 2016. Trajectory generation for quadrotor based systems using numerical optimal control. In *IEEE International Conference on Robotics and Automation*. Stockholm, Sweden, May 2016.
17. Gentili, L., R. Naldi, and L. Marconi. 2008. Modeling and control of VTOL UAVs interacting with the environment. In *IEEE Conference on Decision and Control*, 1231–1236. Cancun, Mexico, December 2008.
18. Hehn, Markus, and Raffaello D'Andrea. 2011. Quadrocopter trajectory generation and control. *18th World Congress of the International Federation of Automatic Control*, vol. 44, 1485–1491. Milano, Italy.

19. Huber, Felix, Konstantin Kondak, Kai Krieger, Dominik Sommer, Marc Schwarzbach, Maximilian Laiacker, Ingo Kossyk, Sven Parusel, Sami Haddadin, and Alin Albu-Schaffer. 2013. First analysis and experiments in aerial manipulation using fully actuated redundant robot arm. In *IEEE/RSJ International Conference on Intelligent Robots and Systems*, 3452–3457. Tokyo, Japan, November 2013.
20. Khalil, Hassan K., and J.W. Grizzle. 1996. *Nonlinear Systems*, vol. 3. Prentice hall New Jersey.
21. Kim, S., S. Choi, and H. J. Kim. 2013. Aerial manipulation using a quadrotor with a two DOF robotic arm. In *IEEE/RSJ International Conference on Intelligent Robots and Systems*, 4990–4995. Tokyo, Japan, November 2013.
22. Kim, S., H. Seo, S. Choi, and H.J. Kim. 2016. Vision-guided aerial manipulation using a multirotor with a robotic arm. *IEEE/ASME Transactions on Mechatronics* 21 (4): 1912–1923.
23. Kondak, K., F. Huber, M. Schwarzbach, M. Laiacker, D. Sommer, M. Bejar, and A. Ollero. 2014. Aerial manipulation robot composed of an autonomous helicopter and a 7 degrees of freedom industrial manipulator. In *IEEE International Conference on Robotics and Automation*, 2107–2112. Hong Kong, China, May 2014.
24. Korpela, Christopher, Matko Orsag, and Paul Oh. 2014. Towards valve turning using a dual-arm aerial manipulator. In *IEEE/RSJ International Conference on Intelligent Robots and Systems*, 3411–3416. Chicago, USA, September 2014.
25. Korpela, C.M., T.W. Danko, and P.Y. Oh. 2011. Designing a system for mobile manipulation from an unmanned aerial vehicle. In *IEEE Conference on Technologies for Practical Robot Applications*, 109–114. Woburn, USA, April 2011.
26. Kuindersma, Scott, Frank Permenter, and Russ Tedrake. 2014. An efficiently solvable quadratic program for stabilizing dynamic locomotion. In *IEEE International Conference on Robotics and Automation*, 2589–2594. Hong Kong, China, May 2014.
27. Lippiello, Vincenzo, and Fabio Ruggiero. 2012. Exploiting redundancy in Cartesian impedance control of UAVs equipped with a robotic arm. In *IEEE/RSJ International Conference on Intelligent Robots and Systems*, 3768–3773. Vilamoura, Portugal, October 2012.
28. Lippiello, Vincenzo, Rafik Mebarki, and Fabio Ruggiero. 2013. Visual coordinated landing of a UAV on a mobile robot manipulator. *IEEE International Symposium on Safety, Security, and Rescue Robotics*, 1–7. Linköping, Sweden.
29. Lippiello, V., J. Cacace, A. Santamaria-Navarro, J. Andrade-Cetto, M.A. Trujillo, Y.R. Esteves, and A. Viguria. 2016. Hybrid visual servoing with hierarchical task composition for aerial manipulation. *IEEE Robotics and Automation Letters*, 1 (1): 259–266. Presented in ICRA'16, January 2016.
30. Lippiello, Vincenzo, and Fabio Ruggiero. 2012. Cartesian impedance control of a UAV with a robotic arm. *IFAC Proceedings Volumes* 45 (22): 704–709.
31. Marconi, L., and R. Naldi. 2012. Control of aerial robots: Hybrid force and position feedback for a ducted fan. *IEEE Control Systems* 32 (4): 43–65.
32. Mebarki, Rafik, and Vincenzo Lippiello. 2014. Image moments-based velocity estimation of UAVs in GPS denied environments. *IEEE International Symposium on Safety, Security, and Rescue Robotics*, 1–6. Toyako-cho, Japan.
33. Mebarki, Rafik, Vincenzo Lippiello, and Bruno Siciliano. 2013. Exploiting image moments for aerial manipulation control. In *ASME Dynamic Systems and Control Conference*. Palo Alto, USA, October 2013.
34. Mebarki, Rafik, Vincenzo Lippiello, and Bruno Siciliano. 2014. Image-based control for dynamically cross-coupled aerial manipulation. In *IEEE/RSJ International Conference on Intelligent Robots and Systems*, 4827–4833. Chicago, USA, September 2014.
35. Mebarki, Rafik, and Vincenzo Lippiello. 2014. Image-based control for aerial manipulation. *Asian Journal of Control* 16 (3): 646–656.
36. Mellinger, Daniel, and Vijay Kumar. 2011. Minimum snap trajectory generation and control for quadrotors. In *IEEE International Conference on Robotics and Automation*, 2520–2525. Shanghai, China, May 2011.
37. Meyer, Johannes, Alexander Sendobry, Stefan Kohlbrecher, Uwe Klingauf, and Oskar Von Stryk. 2012. Comprehensive Simulation of Quadrotor UAVs using ROS and Gazebo. In *Simulation, Modeling, and Programming for Autonomous Robots*, 400–411. Springer.

38. Michael, Nathan, Jonathan Fink, and Vijay Kumar. 2010. Cooperative manipulation and transportation with aerial robots. *Autonomous Robots* 30 (1): 73–86.
39. Nakamura, Yoshihiko. 1990. *Advanced Robotics: Redundancy and Optimization*, 1st ed. Boston, MA, USA: Addison-Wesley Longman Publishing Co., Inc.
40. Nakamura, Yoshihiko, Hideo Hanafusa, and Tsuneo Yoshikawa. 1987. Task-Priority based redundancy control of robot manipulators. *The International Journal of Robotics Research* 6 (2): 3–15.
41. Ollero, A., and K. Kondak. 2012. 10 years in the cooperation of unmanned aerial systems. In *IEEE/RSJ International Conference on Intelligent Robots and Systems*, 5450–5451. Vilamoura, Portugal, October 2012.
42. Orsag, Matko, Christopher Korpela, Miles Pekala, and Paul Oh. 2013. Stability control in aerial manipulation. In *American Control Conference*, 5581–5586. Washington, USA, June 2013.
43. Orsag, Matko, Christopher Korpela, and Oh Paul. 2013. Modeling and control of MM-UAV: Mobile manipulating unmanned aerial vehicle. *Journal of Intelligent & Robotic Systems* 69 (1–4): 227–240.
44. Palunko, Ivana, and Rafael Fierro. 2011. Adaptive control of a quadrotor with dynamic changes in the center of gravity. *IFAC Proceedings Volumes* 44 (1): 2626–2631.
45. Palunko, I., P. Cruz, and R. Fierro. 2012. Agile load transportation: Safe and efficient load manipulation with aerial robots. *IEEE Robotics Automation Magazine* 19 (3): 69–79.
46. Papadopoulo, Théodore, and Manolis I. A. Lourakis. 2000. Estimating the jacobian of the singular value decomposition: Theory and applications. In *European Conference on Computer Vision. Lecture Notes in Computer Science*, vol. 1842, eds. Gerhard Goos, Juris Hartmanis, and Jan van Leeuwen, 554–570. Dublin, Ireland, June 2000.
47. Potschka, Andreas, Christian Kirches, Hans Georg Bock, and Johannes P. Schlöder. Reliable solution of convex quadratic programs with parametric active set methods. *Interdisciplinary Center for Scientific Computing, Heidelberg University technical report*, November 2010.
48. Quigley, Morgan, Ken Conley, Brian Gerkey, Josh Faust, Tully Foote, Jeremy Leibs, Rob Wheeler, and Andrew Y. Ng. 2009. ROS: an open-source robot operating system. In *IEEE International Conference on Robotics and Automation. Workshop on open source software*, vol. 3, 5. Kobe, Japan, May 2009.
49. Rossi, Roberto, Angel Santamaria-Navarro, Juan Andrade-Cetto, and Paolo Rocco. 2017. Trajectory generation for unmanned aerial manipulators through quadratic programming. *IEEE Robotics and Automation Letters*, 2 (2): 389–396. Accepted for presentation in ICRA'17, 2017.
50. Santamaría, Daniel, Francisco Alarcón, Antonio Jiménez, Antidio Viguria, Manuel Béjar, and Aníbal Ollero. 2012. Model-based design, development and validation for UAS critical software. *Journal of Intelligent & Robotic Systems* 65 (1–4): 103–114.
51. Santamaria-Navarro, Angel, Patrick Grosch, Vincenzo Lippiello, Joan Solà, and Juan Andrade-Cetto. Uncalibrated visual servo for unmanned aerial manipulation. *Accepted for publication in the IEEE/ASME Transactions on Mechatronics. To appear*, 2017.
52. Santamaria-Navarro, Angel, Vincenzo Lippiello, and Juan Andrade-Cetto. 2014. Task priority control for aerial manipulation. *IEEE International Symposium on Safety, Security, and Rescue Robotics*, 1–6. Hokkaido, Japan, October 2014.
53. Sreenath, Koushil, Nathan Michael, and Vijay Kumar. 2013. Trajectory generation and control of a quadrotor with a cable-suspended load-A differentially-flat hybrid system. In *IEEE International Conference on Robotics and Automation*, 4888–4895. Karlsruhe, Germany, May 2013.
54. Togai, Masaki. 1986. An application of the singular value decomposition to manipulability and sensitivity of industrial robots. *SIAM Journal on Algebraic Discrete Methods* 7 (2): 315–320.
55. Yoshikawa, T. 1985. Manipulability of robotic mechanisms. *The International Journal of Robotics Research* 4 (2): 3–9.
56. Zanchettin, Andrea Maria, and Paolo Rocco. Reactive motion planning and control for compliant and constraint-based task execution. In *IEEE International Conference on Robotics and Automation*, 2748–2753. Seattle, USA, May 2015.

57. Zanchettin, Andrea Maria, and Paolo Rocco. Robust constraint-based control of robot manipulators: an application to a visual aided grasping task. In *IEEE/RSJ International Conference on Intelligent Robots and Systems*, 3634–3639. Daejeon, Korea, October 2016.
58. Zhang, Yunong, Dongsheng Guo, Kene Li, and Jun Li. Manipulability-maximizing self-motion planning and control of redundant manipulators with experimental validation. In *International Conference on Mechatronics and Automation*, 1829–1834. Chengdu, China, August 2012.

Chapter 5
Closing Remarks

The work presented in this book constitutes a step towards an integrated framework to address the problem of driving a UAM using visual information, including the robot state estimation and high-level task control laws. In this final chapter we discuss the conclusions derived from each chapter of this book, gathering them into an overall reflection to highlight the more important ideas and concepts.

In Chap. 2 we presented a state estimation approach to compute the odometry of a flying vehicle by means of light-weight sensors which can work at a high rate requiring low computational burden. We investigated a wide range of algorithm variations with different computing and implementation complexities. The result of our experimentation concludes that the effects of all these variations in the estimator design are minimal. This state estimation method allows small-sized multirotors to be driven autonomously and, in turn, enables UAM platforms to navigate without requiring external infrastructure, such as a motion capture system or GPS. To demonstrate this independence on offboard infraestructure, we designed an $\mathscr{SO}(3)$ controller and presented simulation case studies together with real experiments with all algorithms running on-line and on-board in a limited computational unit.

The localization precision achieved with this state estimation method is good enough to enable autonomous navigation. However, UAMs require very precise positioning while performing close interaction or manipulation tasks. This positioning performance can be achieved by means of local detection of the target object using visual information. In Chap. 3 we presented several methods to perform visual control. Specifically we described the classical position-based (PBVS) and image-based (IBVS) visual servo methods, together with an image-based approach designed for poorly calibrated cameras (UIBVS). PBVS and IBVS show convergence problems with a wrong initialization of the camera focal length or when subject to noise. Such errors are transfered to the control law, producing unrealistic camera displacements along the optical axis. The proposed uncalibrated method shows robustness to such large errors in camera focal length because this distance is optimized on-line, thus allowing us to get rid of its dependency in the formulation of the image Jacobian. In all these visual control methods, we assumed a holonomic system with 6DoF (i.e., the camera can move freely in space). Then, in order to use them in a real UAM

A. Santamaria-Navarro et al., *Visual Guidance of Unmanned Aerial Manipulators*, Springer Tracts in Advanced Robotics 125,
https://doi.org/10.1007/978-3-319-96580-2_5

and considering its platform underactuation, we need to kinematically augment the multirotor with at least two DoF (e.g., by means of a serial arm).

In Chap. 4 we presented two UAM configurations with regards to the camera mounting: an *onboard-eye-in-hand*, where the camera is rigidly attached to the UAM platform, and an *eye-in-hand*, where the camera is attached at the arm end effector. In both cases, the UAM becomes a redundant robot and we defined three high-level control laws that take advantage of this redundancy, not only to drive the robot using visual information but to do so whilst accomplishing other secondary tasks. These control laws include two hierarchical formulations; one using full least squares, which specifically considers the non controllable state variables (i.e., the underactuation of the aerial platform), and a second hierarchical approach, which might achieve larger constraint-task reconstruction errors than the full least squares solution but with the advantage that algorithmic singularities arising from conflicting tasks are decoupled from the singularities of the secondary tasks. The third control law is a trajectory generation algorithm that uses quadratic programming. In this case the high priority tasks can be set as problem constraints for the numerical optimization problem, and we can adopt a weighting strategy for the rest of the tasks. These weights can be modified on-line to achive desired task behaviours during the mission. The chapter also includes the formulation of cost functions for several tasks specifically designed for UAM to accomplish navigation and close interaction during manipulation missions. These task definitions include:

- Collision avoidance (e.g., avoidance of obstacles and self-collisions).
- End effector tracking with visual servoing (i.e., hybrid and uncalibrated approaches).
- Platform stabilization (e.g., aligning the arm CoG with the platform gravitational vector, driving the arm to a desired configuration, limiting the quadrotor accelerations, increasing the arm manipulability or reducing the forces on the quadrotor horizontal plane).
- Motion profiles (e.g., forcing minimum joint velocities).

For all control laws and most of the tasks presented, we have demonstrated their viability through extensive simulations and real experiments with UAM robots.

This book contributes with multiple developments to the aerial robotics field, and specifically to navigation and control of unmanned aerial manipulators. Derived from the conclusions presented above, a summary of the contributions obtained in this book is listed in the following.

1. **Odometry estimation using low-cost, light-weight and high-rate visual-inertial-range sensors** (Chap. 2). We have presented and analyzed the use of these sensors within a filtering scheme, obtaining an estimated state which is richer than just odometry, and includes higher derivatives such as velocities and accelerations, all precisely referenced to the gravity direction. These are exploited by a non-linear controller to drive the vehicle in 3D space, showing that the employed sensors are more than sufficient to provide positional autonomy to an

aerial platform. This is the first time that such inexpensive sensors enable precise localization and autonomous navigation of aerial vehicles.

2. **Odometry estimation using light algorithms** (Chap. 2). We have shown the feasibility of using such low-cost sensor setup with light algorithms to achieve not only hovering maneuvers but also fully autonomous navigation. We investigated a wide range of algorithm variations with different computing and implementation complexities, and the result of our experimentation shows that the effects of all the variations in the estimator design are minimal. In particular:
   - We found equivalent performances for EKF and ESKF filter types.
   - The refinements on the IMU transition matrices ($\mathbf{F}_1 \cdots \mathbf{F}_3$) have no conclusive effect, meaning that the classical Euler approximation ($\mathbf{F}_1$) is sufficiently good.
   - The difference between quaternion integrators (Q0B, Q0F and Q1) is minimal.
   - The error composition (LE or GE) has no implications on the estimation results.
3. **The UIBVS** (Chap. 3). We described the principles of position- and image-based visual servoing, together with a new method for uncalibrated cameras. A new image Jacobian is derived without the dependency on the camera focal length while guaranteeing asymptotic stability of the control law regardless of the target point selection, as long as planar configurations are avoided.
4. **Task priority control using hierarchical control laws** (Chap. 4). We presented two different hierarchical control laws that take advantage of UAM DoF redundancy and consider the underactuation of their platforms. A first formulation is based on a full least squares solution, whereas a second one presents a similar approach but this time decoupling algorithmic singularities between tasks. In both cases, we can tackle additional objectives expressed as constraints (i.e., subtasks). Although hierarchical task composition techniques are well known for redundant manipulators, its use on aerial manipulation is novel.
5. **Trajectory generation using quadratic programming** (Chap. 4). We described an optimization method to generate feasible trajectories for all the joints of a UAM, taking advantage of a quadratic programming approach to optimize robot joint commands to accomplish several tasks. The numerical optimization problem is designed to run on-line and on board a limited computational unit, and allows us to set bounds and constraints. Hence, we can set the critical objectives (e.g., collision avoidance) as a constraint for all other tasks.
6. **UAM task definitions** (Chap. 4). We defined several tasks specifically designed for UAM to accomplish navigation and close interaction phases during manipulation missions. Moreover, we show how the platform non-controllable DoF must be considered in their designs.

Finally, we discuss some of the future research lines that arise from the work and contributions presented above.

Extension of the odometry estimation method: The sensors used in the state estimation approaches presented in Chap. 2 constitutes a minimal sensor suite to obtain the vehicle position, velocity and acceleration. However, some of the states are not observable (i.e., the platform xy position and yaw orientation). Although the drift of the estimated state is small (a few centimeters after one-minute flight) we can think about adding measurement updates with other sensors, such as a 3D compass (in case of a scenario with small earth magnetic field alterations) or partial GPS pseudo-range measurements.

Relative positioning with respect to a target without a prioriy knowledge of it: The target localization has not been the scope of this book, and thus its detection has been simplified by using either an artificial marker or a motion capture system. However, in real applications these methods could not be feasible, requiring a relative positioning with respect to an object without a priori knowledge of it. Hence, it could be interesting to explore methods to obtain this positioning from unknown and unstructured scenarios.

Hierarchical control law improvement: The activation and deactivation of the safety task as well as a dynamic exchange of task priority roles can induce some chattering phenomena, which can be avoided by introducing a hysteresis scheme. Moreover, the dimensionality of the subspace associated to each null space projector is a necessary condition to be considered when designing subtasks, however it might not be sufficient to guarantee the fulfilment of the subtask and a thorough analytical study of these spaces would be interesting.

New hierarchical formulation coupled with the platform attitude controller: All presented control laws are acting in a high-level (i.e., their outputs are references to the arm joint and platform attitude controllers). This decoupling can be done due to the different dynamics of each control loop, however, it would be interesting to formulate a hierarchical control law with both systems coupled.

Consider the vehicle and arm dynamics within the control law: The presented methods to drive the UAM are based on the vehicle kinematics. In this book we have not accounted for the platform and arm dynamics because we have considered only UAM navigation tasks, without aggressive maneuvers. However, in case of requiring fast movements or carrying heavy loads during transportation missions, a meticulous dynamics study must be introduced.

Focus on manipulation tasks: In most of the manipulation tasks to be done with a UAM, contact during prolongued time with the arm end effector would be required. In that sense we feel interesting to explore the addition of compliance to the arm actuation and to the control law design, in order to perform contact tasks.

Appendix 5.A Multimedia Material

The contents of the videos referenced through the book are detailed in this appendix, enumerated in the following list. All videos can be found on the Internet, at the multimedia page of the author's website: http://www.angelsantamaria.eu.

1. **High-frequency MAV state estimation using inertial and optical flow measurement units**. This video accompanies Chap. 2 and demonstrates the odometry

estimation method through real MAV flights. First, experiments benchmarking all filter types against ground-truth (i.e., a motion capture system) are presented. Second, the video shows the technique with flights in a GPS-denied outdoor scenario. All flights are done using the hardware setting A described in Sect. 2.10.2. This video is related to [7].

2. **Autonomous navigation of micro aerial vehicles using high-rate and low-cost sensors**. In contrast to Video 1, Video 2 shows results of using the odometry estimation technique described in Chap. 2 using setting B —see 2.10.2). This video supports [6]. In this case, all algorithms are running on-line and on board the MAV, and the estimated state is used to feed the nonlinear controller presented in Sect. 2.9.
3. **Uncalibrated image-based visual servoing**. This video supports the uncalibrated visual servo method described in Sect. 3.6 and accompanies the paper [3]. The technique outperforms calibrated visual servo schemes in situations with noisy calibration parameters and for unexpected changes in the camera zoom. The method's performance is demonstrated both in simulations and in a ROS implementation of a quadrotor servoing task.
4. **Hybrid visual servoing with hierarchical task composition for aerial manipulation**. The video shows real experiments of the hybrid visual servoing technique presented in Sect. 4.7.1. A hierarchical control law is used to position the UAM end effector with a PBVS while keeping the target in the camera FoV with an IBVS. Moreover, two subtasks are active: A task to vertically align the arm CoG with the platform gravitational vector, and an other task to reach desired arm joint positions. The video presents four main maneuvers: two grasping and two plugging operations with bars. This video accompanies the paper [1].
5. **Task priority control for aerial manipulation**. The hierarchical task controller described in Sect. 4.4.2 is validated in this video through simulation case studies as in the experiments Sect. 4.7.2.1. The mission shown consists on servoing the UAM end effector using visual information for an inspection and surveillance task. The related publication is [5].
6. **Uncalibrated visual servo for unmanned aerial manipulation**. This video shows experiments similar to those presented in Video 5, but this time with a real UAM —see experiments Sect. 4.7.2. The effect of adding hierarchically subtasks (i.e., using the control law presented in Sect. 4.4.2) is specifically shown. This material is related with [4].
7. **Trajectory generation for unmanned aerial manipulators through quadratic programming**. In Sect. 4.5 we presented a technique which applies quadratic programming (i.e., the on-line active set strategy) to generate a feasible joint trajectory, in order to accomplish a set of tasks with defined bounds and inequality constraint. This video accompanies the Sect. 4.5 and shows a mission composed of two phases, navigation and interaction, performed by an aerial manipulator. Weights of the cost functions are assigned in order to perform the two phases with different strategies. During the navigation phase, arm joints are guided towards a desired configuration for the sake of stability and motion is performed

mainly with quadrotor DoFs. On the other hand, during the interaction phase, motion is performed by robot arm joints to obtain more accuracy. This video is related with [2].

References

1. Lippiello, V., J. Cacace, A. Santamaria-Navarro, J. Andrade-Cetto, M.A. Trujillo, Y.R. Esteves, and A. Viguria. 2016. Hybrid visual servoing with hierarchical task composition for aerial manipulation. *IEEE Robotics and Automation Letters*, 1(1): 259–266. Presented in ICRA'16.
2. Rossi, Roberto, Angel Santamaria-Navarro, Juan Andrade-Cetto, and Paolo Rocco. 2017. Trajectory generation for unmanned aerial manipulators through quadratic programming. *IEEE Robotics and Automation Letters*, 2(2): 389–396. Accepted for presentation in ICRA'17.
3. Santamaria-Navarro, Angel, and Juan Andrade-Cetto. 2013. Uncalibrated image-based visual servoing. In *IEEE international conference on robotics and automation*, pp. 5247–5252, Karlsruhe, Germany.
4. Santamaria-Navarro, Angel, Patrick Grosch, Vincenzo Lippiello, Joan Solà , and Juan Andrade-Cetto. 2017. Uncalibrated visual servo for unmanned aerial manipulation. *Accepted for publication in the IEEE/ASME Transactions on Mechatronics*. To appear.
5. Santamaria-Navarro, Angel, Vincenzo Lippiello, and Juan Andrade-Cetto. 2014. Task priority control for aerial manipulation. *IEEE International Symposium on Safety*, 1–6., Security, and Rescue Robotics Japan: Hokkaido.
6. Santamaria-Navarro, Angel, Giuseppe Loianno, Joan Solà, Vijay Kumar, and Juan Andrade-Cetto 2017. Autonomous navigation of micro aerial vehicles: State estimation using fast and low-cost sensors. Submitted to Autonomous Robots.
7. Santamaria-Navarro, Angel, Joan Sola, and Juan Andrade-Cetto. 2015. High-frequency MAV state estimation using low-cost inertial and optical flow measurement units. In *IEEE/RSJ international conference on intelligent robots and systems*, pp. 1864–1871, Hamburg, Germany.

The manufacturer's authorised representative in the EU is Springer Nature Customer Service Centre GmbH, Europaplatz 3, 69115 Heidelberg, Germany. If you have any concerns regarding our products, please contact ProductSafety@springernature.com

Printed and bound by CPI Group (UK) Ltd, Croydon, CR0 4YY
30/09/2026
02225664-0017